Cell Surface GRP78, a New Paradigm in Signal Transduction Biology

Cell Surface GRP78, a New Paradigm in Signal Transduction Biology

Edited by

Salvatore V. Pizzo

Duke University Medical Center, Durham, NC, United States

Academic Press is an imprint of Elsevier
125 London Wall, London EC2Y 5AS, United Kingdom
525 B Street, Suite 1800, San Diego, CA 92101-4495, United States
50 Hampshire Street, 5th Floor, Cambridge, MA 02139, United States
The Boulevard, Langford Lane, Kidlington, Oxford OX5 1GB, United Kingdom

Copyright © 2018 Elsevier Inc. All rights reserved.

No part of this publication may be reproduced or transmitted in any form or by any means, electronic or mechanical, including photocopying, recording, or any information storage and retrieval system, without permission in writing from the publisher. Details on how to seek permission, further information about the Publisher's permissions policies and our arrangements with organizations such as the Copyright Clearance Center and the Copyright Licensing Agency, can be found at our website: www.elsevier.com/permissions.

This book and the individual contributions contained in it are protected under copyright by the Publisher (other than as may be noted herein).

Notices

Knowledge and best practice in this field are constantly changing. As new research and experience broaden our understanding, changes in research methods, professional practices, or medical treatment may become necessary.

Practitioners and researchers must always rely on their own experience and knowledge in evaluating and using any information, methods, compounds, or experiments described herein. In using such information or methods they should be mindful of their own safety and the safety of others, including parties for whom they have a professional responsibility.

To the fullest extent of the law, neither the Publisher nor the authors, contributors, or editors, assume any liability for any injury and/or damage to persons or property as a matter of products liability, negligence or otherwise, or from any use or operation of any methods, products, instructions, or ideas contained in the material herein.

British Library Cataloguing-in-Publication Data
A catalogue record for this book is available from the British Library

Library of Congress Cataloging-in-Publication Data
A catalog record for this book is available from the Library of Congress

ISBN: 978-0-12-812351-5

For Information on all Academic Press publications
visit our website at https://www.elsevier.com/books-and-journals

Publisher: John Fedor
Acquisition Editor: Sara Tenney
Editorial Project Manager: Fenton Coulthurst
Production Project Manager: Debasish Ghosh
Cover Designer: MPS

Typeset by MPS Limited, Chennai, India

CONTENTS

LIST OF CONTRIBUTORS

Ali A. Al-Hashimi
St. Joseph's Healthcare Hamilton and McMaster University, Hamilton, ON, Canada

Richard C. Austin
St. Joseph's Healthcare Hamilton and McMaster University, Hamilton, ON, Canada

Robin E. Bachelder
Duke University Medical Center, Durham, NC, United States

Mario Gonzalez-Gronow
Duke University Medical Center, Durham, NC, United States

Udhayakumar Gopal
Duke University Medical Center, Durham, NC, United States

Amy S. Lee
USC Norris Comprehensive Cancer Center, Los Angeles, CA, United States

Salvatore V. Pizzo
Duke University Medical Center, Durham, NC, United States

Janusz Rak
McGill University, Montreal, QC, Canada

Rupa Ray
Duke University Medical Center, Durham, NC, United States

Yuan-Li Tsai
USC Norris Comprehensive Cancer Center, Los Angeles, CA, United States

PREFACE

Why write this book now? At first glance, the topic may seem esoteric and just about one more signaling pathway. Yet this is a most unusual mechanism which repurposes an intracellular protein, the Glucose Regulated Protein of molecular weight 78,000 (GRP78), as a cell surface signaling receptor. There are precedents for such behavior, for example cell surface ATP synthase as the angiostatin receptor. However, cell surface GRP78 has unique properties. The intracellular version of the protein exists in the endoplasmic reticulum (ER) where it is a key player in the unfolded protein response, essential for cell survival under conditions of stress. Cells may be stressed by many factors such as hypoxia, disregulation of glucose metabolism, or injury to tissues requiring rapid division of cells to repair the damage. Under these conditions, protein synthesis ramps up and there is a risk of piling up misfolded proteins in the ER. GRP78 is crucial for driving proper folding of such proteins, and its ATP binding domain is essential for these functions. When GRP78 reaches the cell surface, this ATP binding domain is essential, allowing GRP78 to autophosphorylate and function like a classic tyrosine kinase-type receptor. To our knowledge, this represents unique behavior for a repurposed intracellular protein. Of particular interest, GRP78 as a cell surface receptor is generally associated with disordered cell biology. Other than a small fraction of the population of "activated" macrophages, it does not demonstrate signal transduction properties in normal cells. Cancer cells are the main arena where cell surface GRP78 functions as a pro-proliferative, promigratory, and antiapoptotic receptor. But, similar biology is seen in synovial cells from rheumatoid arthritis patients, and in damaged endothelial cells such as are seen in atherosclerosis. On the cell surface, GRP78 associates with many other proteins playing a regulatory "dance." One such example to be discussed in detail in this book is its association with the procoagulant molecule tissue factor. Indeed, regulation of these complexes is almost certainly critical with respect to the association of cancer and thrombotic events.

Thus, we feel that this is a good time for such a book and we thank Elsevier for the opportunity.

Salvatore V. Pizzo

June 2017

CHAPTER 1

An Historical Perspective: Cell Surface GRP78, a New Paradigm in Signal Transduction Biology

Salvatore V. Pizzo
Duke University Medical Center, Durham, NC, United States

DISCOVERY OF CELL SURFACE GRP78

This book will focus on the presence of the Glucose Regulated Protein 78,000 (GRP78) at the cell surface and its role in signal transduction. GRP78 is best known for its function as a molecular chaperone in the endoplasmic reticulum (ER; see Chapter 3: Cell Surface GRP78: Anchoring and Translocation Mechanisms and Therapeutic Potential in Cancer). As such, its presence on the cell surface was not anticipated. The story of cell surface GRP78 actually begins with the study of the proteinase inhibitor α_2-macroglobulin (α_2M). α_2M is a broad-based proteinase inhibitor that first appeared some 600,000,000 years ago.[1] Its persistence over such a long period of time is somewhat perplexing since, in general, it appears to play a secondary role as an inhibitor of proteinases. Over the course of evolution, many much more specific inhibitors have appeared such as α_1-proteinase inhibitor, antithrombin III, and tissue inhibitors of metalloproteinase. Nevertheless, no known case of a total α_2M deficiency has been identified in humans.[1] These facts led us to propose that α_2M is a sensor of proteolysis, rather than primarily an inhibitor of proteinases.[2] α_2M is ideally suited for such a role, since it binds proteinases from all four mechanistic classes.[1] When activated, a thiolester bond in each of its four identical subunits ruptures and the proteinase undergoes a major conformational change.[3,4] The molecule undergoes a 10% compacting

Cell Surface GRP78, a New Paradigm in Signal Transduction Biology. DOI: https://doi.org/10.1016/B978-0-12-812351-5.00001-5
© 2018 Elsevier Inc. All rights reserved.

of its structure exposing receptor recognition sites in each of its four subunits.[5] The activated form is designated α_2M^* to distinguish it from the "native" molecule, α_2M. It was recognized as early as the 1980s that receptors are present on a variety of cells in the body including macrophages and hepatocytes which cause rapid removal of α_2M^* from blood, $t_{1/2} = 2-5$ min.[5,6] The first receptor identified as an α_2M^* binding site by Strickland and his colleagues was the lipoprotein receptor-related protein (LRP), a broad-based catabolic receptor which removes various proteins from the circulation.[7] We argued that a second, signal transducing receptor must exist to explain how α_2M could function as a sensor of proteolysis.[2] If α_2M^* is to function as such a sensor, cellular mechanism(s) must exist to transduce information to the signaling machinery of the cells which recognize the protein. That is to say, an α_2M^* signaling receptor (α_2M^*SR) must exist. There was one study demonstrating the ability of α_2M^* to stimulate prostaglandin E_2 synthesis,[8] but the search for a signaling receptor began in earnest in 1993. Detailed studies of α_2M^* and signal transduction became possible with the arrival of Dr Uma Misra in the Department of Pathology at the Duke Medical School. Misra was an expert in signal transduction biochemistry. He performed the initial studies employing elicited murine peritoneal macrophages. Misra was able to show that when treated with α_2M^* a percentage of these cells demonstrated increased intracellular calcium, inositol phosphates, and cyclic AMP.[9] He subsequently demonstrated activation of phospholipase C, phospholipase A_2, and protein kinase C in these cells.[10] In 1994 he showed that the α_2M signaling receptor must be distinct from LRP.[11] This work was extended by mutation studies of the carboxyl-terminal domain of α_2M where the receptor binding site is located. Mutation of specific amino acid residues affected binding of the receptor binding domain to either LRP or the as yet unidentified α_2M^*SR.[12] Subsequent studies employing ^{125}I- α_2M^* demonstrated that activated murine peritoneal macrophages have two distinct classes of binding sites as shown by a biphasic Scatchard plot.[13] One site was clearly LRP ($K_d \sim 1-10$ nM, $\sim$250,000 binding sites/cell) while the other was the α_2M^*SR ($K_d \sim 50-100$ pM, $\sim$10,000 binding sites/cell). By contrast, naïve murine peritoneal macrophages possess only LRP.[14] In a number of subsequent publications, Misra dissected the α_2M^*-dependent signaling pathways (e.g., see Refs. 15–26). These pathways will be discussed in greater detail by Dr Udhayakumar Gopal in Chapter 2, The Endoplasmic Reticulum Chaperone GRP78 Also

Functions as a Cell Surface Signaling Receptor. The major breakthrough with respect to identification of this receptor occurred in 2002 when we isolated and characterized the cell surface GRP78 as the α_2M^*SR.[17] We had by this time undertaken studies of α_2M^*-mediated signal transduction in prostate cancer cells as well as activated macrophages (e.g., see Ref. 19).

GRP78 AND CANCER CELL BIOLOGY

Scientific progress often occurs for odd reasons. Dr Misra is a devout Hindu and vegetarian. One day I asked him how he felt about sacrificing mice for his studies. He said that he always prayed for their little souls. This troubled me and (not for the first time) I urged him to switch to cancer cells. Perhaps they might also express α_2M^*SR on the cell surface. This seemed like a logical possibility to me since the pathways activated were pro-proliferative in nature. We had a number of prostate cancer cell lines frozen in storage including PC-3 cells and their derived offspring 1-LN cells. The latter were obtained from a lymph node metastasis. They demonstrate a much more aggressive phenotype than their "parent" cells.

Cancer cells exist in a rather hostile environment. Oxygen is often limited as its vasculature is often poor from a structural standpoint, and the vessels are "leaky." It may also struggle to obtain an adequate supply of glucose to fuel its metabolism. As is well known, most tumors do not efficiently utilize glucose since they depend primarily on glycolysis via the Embden–Meyerhof pathway which produces only 2 moles of ATP per mole of glucose. By contrast, oxidative metabolism produces some 36 additional molecules of ATP per molecule of glucose. Why would a rapidly growing cancer cell function in this way? In Chapter 3, Cell Surface GRP78: Anchoring and Translocation Mechanisms and Therapeutic Potential in Cancer, we will propose one possible reason for this phenomenon first described in the 1920s by Otto Warburg for tumors.

Whatever the reason for this metabolic behavior, cancer cells are clearly under great stress and show upregulation of the Unfolded Protein Response, essential for proper protein folding in the ER when cells are synthesizing an increased level of proteins. Perhaps, the increased synthesis of the key molecular chaperone GRP78 results in the escape of some of this protein from the ER and its arrival on the cell surface? This, at any rate, was our thinking. Within several months, a study appeared demonstrating that the appearance of autoantibodies to GRP78 in the plasma of patients suffering from prostate cancer indicated a very poor prognosis and correlated with an increased probability for metastases.[27] We subsequently confirmed these observations and extended them to patients with melanoma.[28] We were able to identify the binding site for both α_2M^* and the patient-derived autoantibodies to a sequence in the amino-terminal domain of GRP78.[28] In addition, we demonstrated that these autoantibodies functioned as receptor agonists, at least in part explaining why their presence in plasma represented a poor prognostic indicator in cancer patients. Many subsequent studies have demonstrated the presence of these autoantibodies in the plasma of patients with diverse tumors including head and neck, lung, liver, colon, brain, and certain hematopoietic malignancies (as discussed in Refs. 29–31). The next major observation with respect to cell surface GRP78 was that antibodies raised against the carboxyl-terminal domain are antagonistic to the effects of either α_2M^* or the autoantibodies in the plasma of cancer patients. Ligating the amino-terminal domain of GRP78 activates pro-proliferative, promigratory, and antiapoptotic signaling pathways. By contrast ligation of the carboxyl-terminal domain of GRP78 activates antiproliferative, antimigratory, and proapoptotic signaling pathways.[32–34] It is evident that both ends of GRP78 must be exposed on the cell surface to explain these effects, and this is in fact the case.[28] Thus, cell surface GRP78 must be in complex with transmembrane binding partners, one of which MTJ-1 has been identified.[22] Indeed, it is possible to isolate a multiprotein signaling complex from prostate cancer cells consisting of GRP78, MTJ-1, SoS, GrB, and Gαq11—the latter being a pertussis toxin-insensitive G-protein.[26] To our knowledge, this is the only known protein receptor with two distinct binding domains demonstrating completely opposite biological effects.

A MOST COMPLEX SIGNALING MECHANISM

The complexity of GRP78-dependent signal transduction deserves further consideration. Our studies demonstrated that ligation of the carboxyl-terminal domain by antibodies always overcomes effects of ligating the amino-terminal domain.[32–34] This is true whether the agonist binding to the amino-terminal domain is α_2M^* or antibody. These observations are the basis for the potential of antibodies directed against the carboxyl-terminal domain to function as an immunotherapeutic. One not so obvious question is whether both the amino- and carboxyl-terminal domain ligands bind to the same GRP78 molecule? This seems unlikely to me since the signaling machinery involved in the events following ligation of either end of the molecule are entirely distinct.[26,32–34] For example, no G protein seems to be required for activation of the pathway which depends on ligation of the carboxyl-terminal domain. GRP78 is a promiscuous binding partner. It is associated with MHC Class I molecules, crypto, and tissue factor among the known binding partners (see Refs. 30,31, Chapter 2: The Endoplasmic Reticulum Chaperone GRP78 Also Functions as a Cell Surface Signaling Receptor and Chapter 5: Novel Cell Surface Targets for the Plasminogen Activating System in the Brain: Implications for Human Disease). These interactions almost certainly modulate the biological behavior seen when the carboxyl-terminal domain is ligated. There are also many ligands for cell surface GRP78—as will be discussed in Chapter 2, The Endoplasmic Reticulum Chaperone GRP78 Also Functions as a Cell Surface Signaling Receptor. The data therefore, suggest that there are multiple populations of cell surface GRP78 molecules. Some are associated with the pro-proliferative signaling machinery which activate Gαq11, and others with the antiproliferative signaling system which requires p53 or other members of this family of macromolecules.[32–34] While such behavior seems unprecedented, it would best fit the observed behavior of cell surface GRP78.

What follows is a more detailed review of the role(s) of cell surface GRP78. It will become evident that α_2M^* binding to GRP78 regulates cancer cell metabolism in a manner more like a growth factor than a proteinase inhibitor.

REFERENCES

1. Hart JP, Pizzo SV. Proteinase inhibitors: α-macroglobulins and kunins. In: Colman RW, Clowes AW, Goldhaber SZ, Marder VJ, George JN, editors. Hemostasis and thromboses: basic principles and clinical practice. 5th ed. Philadelphia: Lippincott, Williams & Wilkins; 2006. p. 395–407 [chapter 21].

2. Chu CT, Howard GC, Misra UK, Pizzo SV. α_2-Macroglobulin: a sensor for proteolysis. *Ann N Y Acad Sci* 1994;**737**:291–307.

3. Gonias SL, Reynolds JA, Pizzo SV. Physical properties of human α_2-macroglobulin following reaction with methylamine and trypsin. *Biochim Biophys Acta* 1982;**705**:306–14.

4. Gonias SL, Pizzo SV. Conformation and protease binding activity of binary and ternary human α_2-macroglobulin-protease complexes. *J Biol Chem* 1983;**258**:14682–5.

5. Imber MJ, Pizzo SV. Clearance and binding of two electrophoretic "fast" forms of human α_2-macroglobulin. *J Biol Chem* 1981;**256**:8134–9.

6. Gonias SL, Balber AE, Hubbard WJ, Pizzo SV. Ligand binding, conformational change and plasma elimination of human, mouse and rat α-macroglobulin proteinase inhibitors. *Biochem J* 1983;**209**:99–105.

7. Herz J, Strickland D. LRP: a multifunctional scavenger and signaling receptor. *J Clin Invest* 2001;**108**:779–84.

8. Hoffman MR, Pizzo SV, Weinberg JB. α_2-macroglobulin-proteinase complexes stimulate prostaglandin E_2 synthesis by peritoneal macrophages. *Agents Actions* 1988;**25**:360–8.

9. Misra UK, Chu CT, Rubenstein DS, Gawdi G, Pizzo SV. Receptor-recognized α-macroglobulin-methylamine elevates intracellular calcium, inositol phosphates, and cyclic AMP in murine peritoneal macrophages. *Biochem J* 1993;**290**:885–91.

10. Misra UK, Pizzo SV. Ligation of α_2M receptors with α_2M-methylamine stimulates the activities of phospholipase C, phospholipase A_2 and protein kinase C in murine peritoneal macrophages. *Ann N Y Acad Sci* 1994;**737**:486–9.

11. Misra UK, Chu CT, Gawdi G, Pizzo SV. The relationship between low density lipoprotein-receptor related protein/$\alpha_2$macroglobulin (α_2M) receptors and the newly described α_2M signaling receptor. *J Biol Chem* 1994;**269**:18303–6.

12. Howard GC, Yamaguchi Y, Misra UK, Gawdi G, Nelsen A, DeCamp DL, et al. Selective mutations in cloned and expressed α_2-macroglobulin receptor binding fragment alter binding to either the α_2-macroglobulin signaling receptor or the low density lipoprotein receptor-related protein/α_2-macroglobulin. *J Biol Chem* 1996;**271**:14105–11.

13. Odom A, Misra UK, Pizzo SV. Nickel inhibits binding of α_2-macroglobulin-methylamine to the low density lipoprotein receptor related protein/α_2-macroglobulin receptor but not the α_2-macroglobulin signaling receptor: evidence for a nickel effect on a region of α_2-macroglobulin upstream of the receptor binding domain. *Biochemistry* 1997;**36**:12395–9.

14. Bhattacharjee G, Misra UK, Gawdi G, Cianciolo G, Pizzo SV. Inducible expression of the α_2-macroglobulin signaling receptor in response to antigenic stimulation: a study of second messenger generation. *J Cell Biochem* 2001;**82**:260–70.

15. Misra UK, Pizzo SV. Regulation of cytosolic phospholipase A_2 activity in macrophages stimulated with receptor-recognized forms of α_2-macroglobulin: role in mitogenesis and cell proliferation. *J Biol Chem* 2002;**277**:4069–78.

16. Misra UK, Akabani G, Pizzo SV. The role of cAMP-dependent signaling in α_2M*-induced cellular proliferation. *J Biol Chem* 2002;**277**:36509–20.

17. Misra UK, Gonzalez-Gronow M, Gawdi G, Hart JP, Johnson CE, Pizzo SV. The role of Grp78 in α_2-macroglobulin-induced signal transduction: evidence from RNA interference that the low density lipoprotein-receptor related protein is associated with, but not necessary for, Grp78-mediated signal transduction. *J Biol Chem* 2002;**277**:42082–7.

18. Misra UK, Gawdi G, Pizzo SV. Induction of mitogenic signaling in the 1LN prostate cell line on exposure to submicromolar concentrations of cadmium. *Cell Signal* 2003;**15**:1059–70.

19. Misra UK, Pizzo SV. Potentiation of signal transduction, mitogenesis and cellular proliferation upon binding of receptor-recognized forms of α_2-macroglobulin to 1-LN prostate cancer cells. *Cell Signal* 2004;**16**:487–96.

20. Misra UK, Gonzalez-Gronow M, Gawdi G, Wang F, Pizzo SV. A novel receptor function for the heat shock protein GRP78: silencing of GRP78 gene expression attenuates α_2M*-induced signalling. *Cell Signal* 2004;**16**:929–38.

21. Misra UK, Pizzo SV. Activation of Akt/PDK signaling in macrophages upon binding of receptor-recognized forms of α_2-macroglobulin to its cellular receptor: effect of silencing the CREB gene. *J Cell Biochem* 2004;**93**:1020–32.

22. Misra UK, Gonzalez-Gronow M, Gawdi G, Pizzo SV. The role of MTJ-1 in cell surface translocation of GRP78, a receptor for α_2-macroglobulin-dependent signaling. *J Immunol* 2005;**174**:2092–7.

23. Gonzalez-Gronow M, Misra UK, Gawdi G, Pizzo SV. Association of plasminogen with dipeptidyl peptidase IV and Na^{+}-H^{+} exchanger isoform NHE_3 regulates invasion of human 1-LN prostate tumor cells. *J Biol Chem* 2005;**280**:27173–8.

24. Misra UK, Sharma T, Pizzo SV. Ligation of cell surface-associated GRP78 by receptor-recognized forms of α_2-macroglobulin activation of PAK-2-dependent signaling in murine peritoneal macrophages. *J Immunol* 2005;**175**:2525–33.

25. Misra UK, Deedwania R, Pizzo SV. Activation and cross talk between Akt, NFκB, and unfolded protein response signaling in 1-LN prostate cancer cells consequent to ligation of cell surface-associated GRP78. *J Biol Chem* 2006;**281**:13694–707.

26. Misra UK, Pizzo SV. Heterotrimeric Gαq11 co-immunoprecipitates with surface-anchored GRP78 from plasma membranes of α2M*-stimulated macrophages. *J Cell Biochem* 2008;**104**:96–104.

27. Mintz PJ, Kim J, Do KA, Wang X, Zinner RG, Cristofanilli M, et al. Fingerprinting the circulating repertoire of antibodies from cancer patients. *Nat Biotechnol* 2003;**21**:57–63.

28. Gonzalez-Gronow M, Cuchacovich M, Llanos C, Urzua C, Gawdi G, Pizzo SV. Prostate cancer cell proliferation *In Vitro* is modulated by antibodies against glucose-regulated protein 78 isolated from patient serum. *Cancer Res* 2006;**66**:11424–31.

29. Pizzo SV. When is a proteinase inhibitor a hormone? The strange tale of α_2-macroglobulin. *J Nat Sci* 2015;**1**(7):e128.

30. Quinones Q, de Ridder GG, Pizzo SV. GRP78: a chaperone with diverse roles beyond the endoplasmic reticulum. *Histol Histopathol* 2008;**23**:1409–16.

31. Gonzalez-Gronow M, Pizzo SV, Misra UK. *GRP78 (BiP): a multifunctional cell surface receptor. Cellular trafficking of cell stress proteins in health and disease.* Dordrecht: Springer Science Business Media; 2013.

32. Misra UK, Mowery Y, Kaczowka S, Pizzo SV. Ligation of cancer cell surface GRP78 with antibodies directed against its carboxyl terminal domain upregulate p53 activity and promote apoptosis. *Mol Cancer Ther* 2009;**8**:1350–62.

33. Misra UK, Pizzo SV. Modulation of the unfolded protein response in prostate cancer cells by antibody-directed against the carboxyl-terminal domain of GRP78. *Apoptosis* 2010;**15**(2):173–82.

34. Misra UK, Pizzo SV. Ligation of cell surface GRP78 with antibody directed against the COOH-terminal domain of GRP78 suppresses Ras/MAPK and PI 3-kinase/Akt signaling while promoting caspase activation in human prostate cancer cells. *Cancer Biol Ther* 2010;**9**:142–52.

CHAPTER 2

The Endoplasmic Reticulum Chaperone GRP78 Also Functions as a Cell Surface Signaling Receptor

Udhayakumar Gopal and Salvatore V. Pizzo
Duke University Medical Center, Durham, NC, United States

Cell Surface GRP78, a New Paradigm in Signal Transduction Biology. DOI: https://doi.org/10.1016/B978-0-12-812351-5.00002-7
© 2018 Elsevier Inc. All rights reserved.

INTRODUCTION

The Glucose Regulated Protein 78,000 (GRP78)—also referred as BiP or HSPA5—is a member of the Heat Shock Protein 70 (HSP70) superfamily and is evolutionarily conserved from yeast to human.[1,2] GRP78 regulates the balance between cancer cell viability and apoptosis by sustaining endoplasmic reticulum (ER) protein folding capacity, by maintaining ER stress sensors, and the ER-associated proapoptotic machineries in their inactive state.[3] Furthermore GRP78, which is traditionally thought to exclusively reside in the ER lumen, can be actively translocated to other cellular locations and be secreted where it has additional functions that control signaling, proliferation, invasion, apoptosis, inflammation, and immunity.[4–7] Cell surface localization of GRP78 was reported in 1997, after which evidence has rapidly accumulated that GRP78 exists on the cell surface of select cell types and functions as a multifunctional receptor.[8] More importantly, cell surface-associated GRP78 is primarily observed in pathological conditions such as cancer. Its presence is associated with resistance to chemotherapy as well as stress conditions, including hypoxia, glucose starvation, radiation, and chemotherapy. This suggests potential applications in targeted therapy as well as applications as a prognostic marker. In this chapter, we summarize the functional importance of cell surface GRP78 (CS-GRP78) and its influence on fundamental biological processes.

RELEASE OF CS-GRP78 INTO THE EXTRACELLULAR ENVIRONMENT

The mechanisms of GRP78 translocation from the ER to the plasma membrane are just emerging, but it may involve oversaturation of the ER retention system, cotrafficking with cell surface client proteins, ER transmembrane GRP78 cycling to the cell surface, as well as a specific mechanism employed by tumor cells. Considering the significance of CS-GRP78 from both the basic cell biology and therapeutic targeting perspective, it is important to understand how GRP78 exists stably on the cell surface and how it reaches the cell surface. This is particularly intriguing because the primary amino acid sequence of the mature GRP78 contains only a few weak hydrophobic domains and GRP78 containing the intact KDEL ER retrieval motif is capable of localizing on the cell surface.[9,10] Global profiling of the cell surface proteome of tumor cells clearly reveals the relative abundance of cytosolic heat shock and ER lumen chaperones including GRP78,[11] suggesting that relocating these stress-inducible chaperones to the cell surface could represent a common adaptive mechanism for cells to respond to stress perturbing protein homeostasis.

The COOH-terminal tetra-peptide KDEL prevents GRP78 secretion and maintains it within the ER lumen.[12] Since KDEL receptor expression is not coordinately upregulated with ER stress in HeLa cells,[13] increases in GRP78 triggered by ER stress may exceed the retention capacity of the KDEL retrieval system which results in an escape from the ER to the cell surface. It is also possible that the activity of the various components of the KDEL system is altered under ER stress or pathological conditions. Another possible mechanism for GRP78 transport to the cell surface may involve the masking of the KDEL motif by glycosylation or other modifications to the protein sequence adjacent to KDEL. A glycosylated form of GRP78 has been detected and potential glycosylation sites exist at the COOH-terminus in close proximity to the KDEL motif.[9,14,15] Additionally, specific GRP78-interacting protein partners may facilitate its transport from the ER to the cell surface and this may be cell type-specific. For example, a DnaJ-like transmembrane protein, MTJ-1, binds GRP78 while silencing MTJ-1 expression apparently suppresses CS-GRP78 expression in macrophages.[16] However, in PC-3 cells, Par-4 is required for translocation of GRP78 from the ER to the plasma membrane.[17]

SECRETED GRP78

In exocrine pancreatic cells, immunogold electron microscopy studies demonstrate that GRP78 is exported from the ER to other intracellular organelles and is even secreted into the extracellular space.[18] Since all the chaperone proteins have the KDEL ER-retention signal, it is possible that saturation of KDEL receptors or defects in the protein sorting system might cause an inability to retrieve these KDEL-bearing proteins for retention in the ER lumen, especially in cells with intensive requirements for protein synthesis and maturation. After these findings, secretion of GRP78 was also detected in tumors. While treating the tumor cell lines with bortezomib, a proteasome inhibitor of tumor angiogenesis, it was discovered that a few of the cell lines secreted significant amounts of GRP78 into the tumor microenvironment.[19] Moreover, tumor cells under ER stress secrete GRP78 which binds to endothelial cell-surface receptors to activate the ERK and AKT pathways and protects endothelial cells from the antiangiogenic effect of bortezomib.[19] In a proteomic study of gastric cancer, GRP78 was identified in the sera of 28% of patients, but not in healthy individuals.[20] Importantly, circulating GRP78 autoantibodies have been detected in patients with gastric and prostate cancer and are implicated in tumor cell proliferation.[20,21] Interestingly, GRP78 was also detected in oviductal fluids from women in the periovulatory period which has the ability to modulate sperm-zona pellucida binding during fertilization.[22] Therefore, secreted GRP78 can potentially regulate a multitude of biological processes in both pathological and physiological conditions.

GRP78 IN EXOSOMES

There is emerging evidence that tumor and tumor stromal cells release exosomes that participate in local and systemic cell communication in an autocrine and paracrine manner. Exosomes have a topology similar to cells and contain a wide range of biologically active materials including cytokines, growth factors, extracellular matrix (ECM) molecules, mRNAs, and microRNAS.[23] Exosomes containing RNA molecules may act as a vehicle for the horizontal transfer of genetic information between the cells.[24] GRP78 has been reported in the exosomes of ovarian cancer patients and it may serve as a diagnostic marker.[25] GRP78 is also expressed in tumor-derived melanoma exosomes.[26] A recent study demonstrates that GRP78 can also be secreted into the extracellular space via exosomes in colon cancer.[27]

INDUCTION OF GRP78 TO THE CELL SURFACE

As a stress response protein, GRP78 expression on cancer cells is potentiated under hypoxic conditions in which tumor angiogenesis, glucose metabolism, and invasion are supported by activating the hypoxia inducible factor 1 signaling pathway.[28] Our laboratory previously showed that in prostate cancer cells activated α_2-macroglobulin (α_2M^*) promotes the transcriptional and translational upregulation of TFII-1 which induces the CS-GRP78 by two- to threefold.[29]

Pretreatment of PC-3 cells with the TRAIL proapoptotic protein or with ER-stress inducers, such as thapsigargin or tunicamycin also induces expression of GRP78 at the plasma membrane.[30–32] GRP78 induction plays a critical role in maintaining cell viability against several kinds of stress including ER Ca^{2+} depletion and accumulation of unglycosylated proteins. ER stress can be chemically induced by agents that inhibit the glycosylation of newly synthesized proteins or by agents that induce ER Ca^{2+} depletion.[31,33,34] One study using prostate cancer cell lines, demonstrated that both the calcium ionophore ionomycin and the glycosylation inhibitor tunicamycin induces GRP78 expression.[31] However, induction of GRP78 and apoptosis was only observed in response to ionomycin, which produces ER Ca^{2+} depletion, but not in response to inhibition of protein glycosylation, suggesting the presence of at least two different pathways mediating the GRP78 stress response. Thapsigargin, an inhibitor of the ubiquitous sarcoendoplasmic reticulum Ca^{2+} ATPase, also induces the expression of GRP78.[30] Inhibition of the reticulum Ca^{2+}-ATPases leads to the depletion of intraluminal Ca^{2+} and a concurrent increase of cytosolic free calcium Ca^{2+}, which plays a critical role in the TG-induced ER stress mechanism.[35] VEGF can induce cell surface expression of GRP78 in endothelial cells and knockdown of GRP78 suppresses VEGF-mediated MAPK signaling and endothelial cell proliferation suggesting feedback regulation.[36] Alpha-Synuclein is a cytosolic protein which is released from neurons. Extracellular synuclein induces an increase in CS-GRP78 which becomes clustered in micro domains of the neuronal plasma membrane.[37]

Preclinical studies indicated that chemotherapy with doxorubicin induces CS-GRP78 in breast cancer and indicates as a potential marker for response to chemotherapy.[38] According to our model, chemotherapy targets the fast proliferating tumor bulk cells whereas the antiproliferative C38 monoclonal antibody which binds to the COOH-terminal domain (to be discussed later) will target the cancer stem cells. It suggests that the combination of GRP78 antibodies with chemotherapy increases the efficacy of ovarian cancer treatment.[39] Moreover, some studies show that XRT triggers the induction of CS-GRP78 in glioblastoma multiforme and nonsmall cell lung cancer, whereas anti-GRP78 antibodies enhanced the efficacy of radiotherapy both in vitro and in vivo.[40] These studies suggest that GRP78 is a promising novel target and anti-GRP78 antibodies could be used as an effective cancer therapy alone, or in combination with radiation or chemotherapy.

CS-GRP78 LIGANDS AND THEIR CELLULAR EFFECTS

Ligands that activate the CS-GRP78 signaling pathway can have variable and even opposing effects on cellular proliferation, apoptosis and cell survival as well as metabolism (Fig. 2.1). Cell surface GRP78, a high-affinity receptor for α_2M^* promotes proliferation, survival, and metastases of prostate cancer cells. In patients with prostate cancer, α_2M^* is proteolytically activated and signals predominately through GRP78 to promote the proliferation and survival of cancer cells.[5] Another study demonstrates that ligation of CS-GRP78 by α_2M^* upregulates the prostate-specific antigen (PSA) which is secreted into the medium and binds α_2M^*. The resultant α_2M^*/PSA complexes interacts with CS-GRP78 to promote more aggressive behavior of prostate cancer by regulating DNA and protein synthesis.[41] Isthmin (ISM) is a secreted 60-kDa protein and further studies suggest that ISM is a novel proapoptotic ligand for CS-GRP78 triggering apoptosis in both cancer cells and cancer endothelial cells by inducing mitochondrial dysfunction.[42] Therefore, targeting the ISM-CS-GRP78 interaction has significant potential as a cancer specific and dual-targeting anticancer agent. Plasminogen Kringle 5 also induces apoptosis of dermal microvessel endothelial cells in a manner that requires CS-GRP78.[43] These data indicate GRP78 as a promising target for cancer therapy. Prostate apoptosis response-4 (Par-4) is a proapoptotic protein which interacts with CS-GRP78 to induce apoptosis via ER stress and activation of the FADD/caspase-8/caspase-3 pathway. Moreover, apoptosis inducible by TRAIL, which also exerts cancer cell-specific effects, is dependent on the Par-4/CS-GRP78 signaling pathway for the induction of apoptosis.[17,44–46] These studies suggest that Par-4 and Kringle 5 are both ligands which interact with CS-GRP78 to trigger apoptosis whereas α_2M^* as a ligand leads to enhancement of cell survival and proliferation.

Vaspin is a novel ligand for the CS-GRP78/MTJ-1 complex and its subsequent signals exert beneficial effects on ER stress-induced metabolic dysfunctions.[47] MICA is a ligand of the activating receptor NKG2D expressed by NK and T cells. MICA interacts with CS-GRP78 which regulates immune evasion in chronic lymphocytic

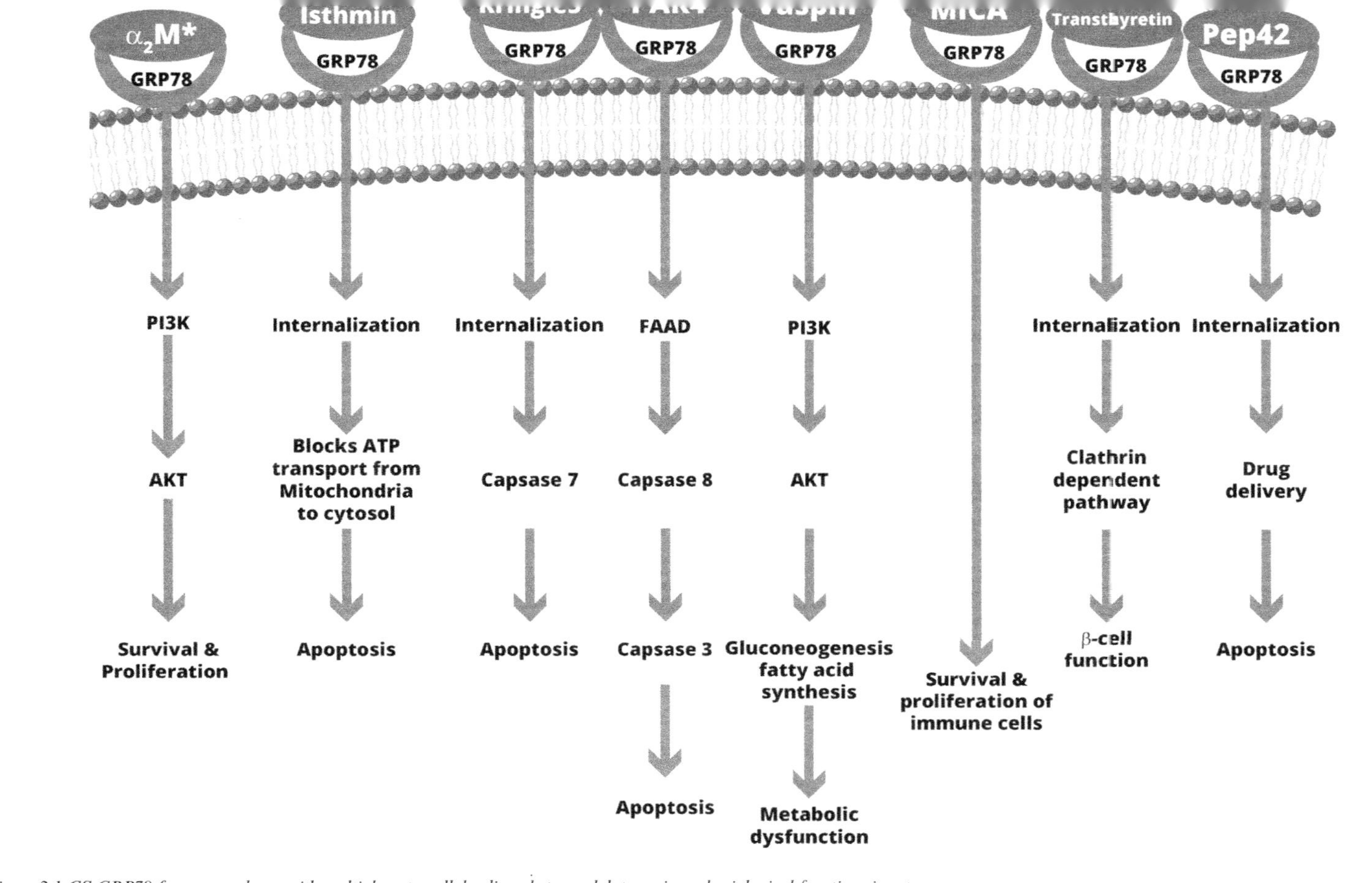

Figure 2.1 CS-GRP78 forms complexes with multiple extracellular ligands to modulate various physiological functions in a tumor.

leukemia.[48] Transthyretin is a functional protein in the pancreatic β-cell. Its interaction with CS-GRP78 regulates its internalization, a process which affects β-cell function.[49] Pep42 a cyclic oligopeptide specifically interacts with CS-GRP78 and the complex is internalized into cells enabling taxol-conjugated Pep42 to target and kill melanoma cells by recognizing GRP78 on the cell surface.[50] Further studies have shown that the fusion of Pep42 conjugated to apoptosis-inducing peptide D-(KLAKLAK)2 selectively kills human cancer cell lines in vitro by binding to CS-GRP78, but with minimal toxicity to normal cells where no GRP78 was detected on the cell surface.[51]

CS-GRP78 FORMS COMPLEXES WITH A VARIETY OF CELL-SURFACE ANCHORED PROTEINS

GRP78-interacting protein partners may facilitate its transport from the ER to the cell surface, and this may be cell type-specific and/or acting in combination. For example, a DnaJ-like transmembrane protein, MTJ-1, binds GRP78 and silencing MTJ-1 expression apparently suppresses cell-surface GRP78 expression in macrophages[16] (Fig. 2.2). GRP78 on the cell surface inhibits tissue factor procoagulant (TF) activity via physical interaction between the TF extracellular domain and a region localized in the COOH-terminal domain of GRP78, distant from either its ATP or peptide binding domain.[52] Furthermore, anti-GRP78 autoantibodies to CS-GRP78 increase TF-procoagulant activity (PCA) through a mechanism that involves the release of Ca^{2+} from ER stores. Therefore blocking GRP78 signaling on the surface of cancer cells attenuates TF-PCA and has the potential to reduce the risk of cancer-related venous thromboembolism[53,54] (Fig. 2.2). This topic will be discussed in detail in Chapter 4, Cell Surface GRP78: A Novel Regulator of Tissue Factor Procoagulant Activity.

CS-GRP78 interacts with Low density lipoprotein receptor-related protein (LRP1), which also binds α_2M^*, and this may modulate α_2M^*-mediated signal transduction pathways. Interestingly, the interaction between GRP78 and LRP1 is greater in cancer cells when compared to macrophages.[55] Moreover, CS-GRP78 interaction with Kringle 5 induces apoptosis that can be aggravated by radiation, which induces internalization of CS-GRP78 by LRP1, resulting in the activation of p38 MAPK[56] (Fig. 2.2). In colon cancer cells, CS-GRP78 interacts with integrin-β1 to regulate colon cancer metastasis by modulating

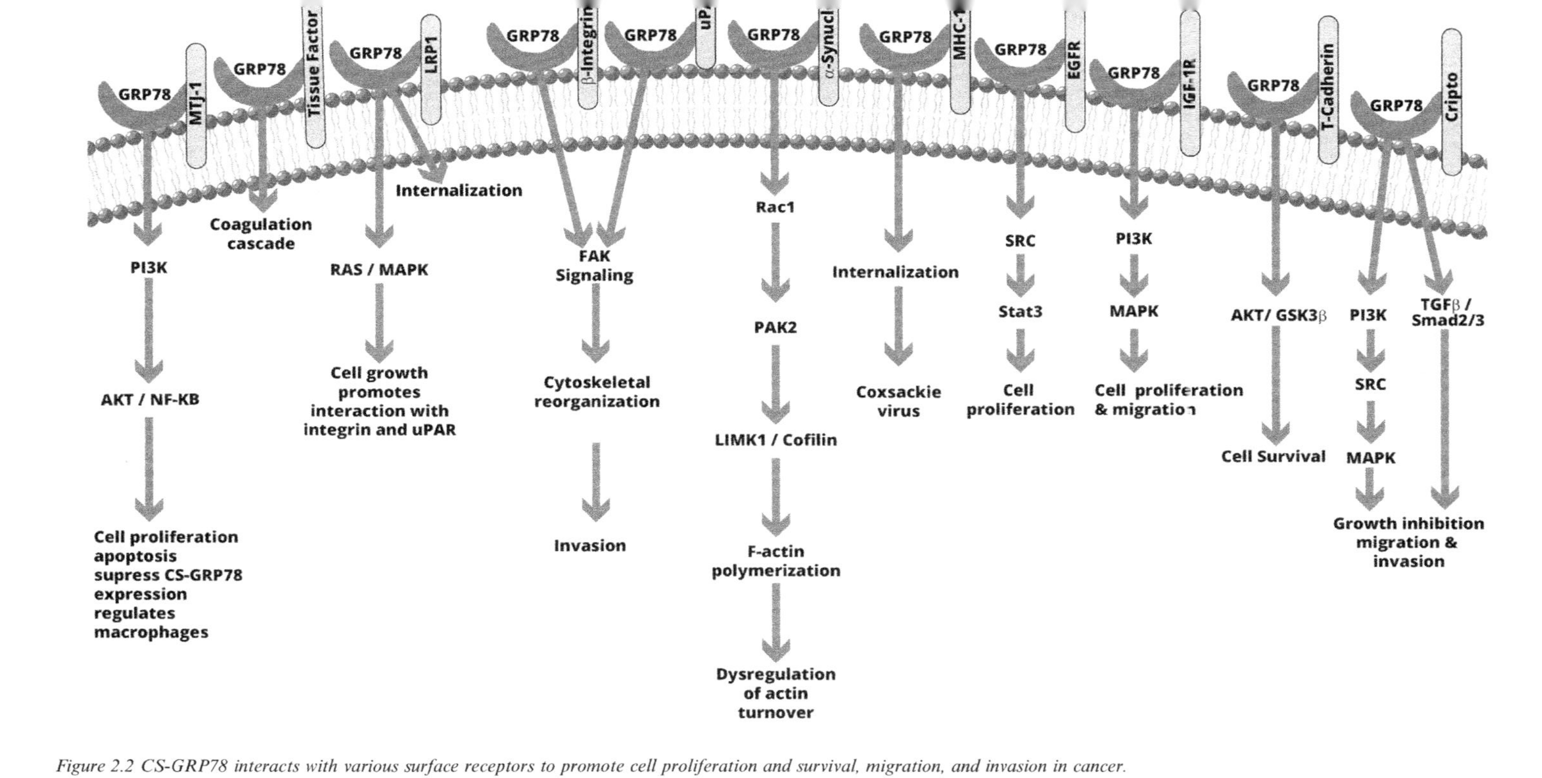

Figure 2.2 CS-GRP78 interacts with various surface receptors to promote cell proliferation and survival, migration, and invasion in cancer.

cell matrix adhesion through focal adhesion kinase (FAK).[57] CS-GRP78 can also interact with the urokinase-type plasminogen activator (uPA)-urokinase plasminogen activator surface receptor (uPAR) protease system to facilitate ECM degradation and promote cell invasion.[57] CS-GRP78 promotes ColoRectal Cancer (CRC) cell migration and invasion by regulating cell matrix adhesion and ECM degradation which is independent of its signaling receptor function. CS-GRP78 interacts with alpha-synuclein which activates signaling cascade leading to cofilin-1 inactivation and stabilization of microfilaments, thus affecting morphology and dynamics of actin cytoskeleton in neurons which leads to neurodegeneration[37] (Fig. 2.2).

GRP78 associates with Major histocompatibility complex (MHC) class I on the surface of these cells and is required for MHC class I expression.[58] Since GRP78 controls the folding of MHC-1,[59] its association involves a site in the peptide binding domain, near the COOH-terminus of GRP78. The complex between these two proteins serves as a receptor for coxsackievirus A9, which interacts only with GRP78, followed by internalization via MHC-1-associated endocytosis[58] (Fig. 2.2). Since MHC class I is ubiquitous in its expression, one would expect GRP78 localization to the cell surface of most cells. Yet, it is not detected on most normal cells by ligand binding or flow studies employing specific antibodies to GRP78. We suggest that GRP78 in complex with MHC class I exists in a different conformation than occurs in the various pathologic conditions we have described. It clearly is not available to bind its ligand α_2M^* in normal cells, and such cells do not signal when exposed to α_2M^*. Thus, GRP78 may actually be present on the surface of most cells, but the NH_2-terminal domain α_2M^* binding site is unavailable, either because of the presence of a binding partner or an altered GRP78 conformational state. In a hepatocellular carcinoma, secreted GRP78 interacts with the EGFR tyrosine kinase to activate cancer cells through EGFR-SRC pathways[60] (Fig. 2.2). CS-GRP78 also interacts with IGF-1R tyrosine kinase to regulate its downstream signaling components including PI3K and MAPK pathways for cellular proliferation and migration[61] (Fig. 2.2). Glycosylphosphatidylinositol (GPI) anchored T-cadherin is reported to associate with GRP78 on the surface of vascular endothelial cells and, in this capacity, GRP78 influences endothelial cell survival as a cell-surface signaling receptor. The prosurvival effects of GRP78 in the unfolded protein cascade and its association with overexpressed T-cadherin in the cardiovasculature and with cardiomyocytes, suggests that GRP78 plays a central role in vascular tissue remodeling and stress[62] (Fig. 2.2). CS-GRP78 interacting with another GPI-anchored protein Cripto—also known as teratocarcinoma-derived growth factor 1—is a developmentally regulated oncoprotein.[63] Disruption of the cell surface GRP78-Cripto complex blocked Cripto activation of MAPK and PI3K pathways and blocked Cripto modulation of activin A, activin B, nodal, and transforming growth factor-β1 (TGFβ1) signaling.[64] Thus, CS-GRP78 is a necessary mediator of Cripto proliferative signaling in human cancer. Moreover, further findings indicate CS-GRP78/Cripto signaling beyond proliferation by demonstrating that it has pivotal and functional roles in the acquisition and maintenance of an invasive, metastatic phenotype in human prostate cancer[65] (Fig. 2.2). Therefore, targeting CS-GRP78/Cripto complex may prevent progression of prostate cancer toward the castration-resistant and metastatic stage.

CS-GRP78 CONTROLS KEY ONCOGENIC SIGNALING PATHWAYS

CS-GRP78 acts as a multifunctional receptor that affects both cell proliferation and viability[4–7] (Fig. 2.3). CS-GRP78 colocalizes with PI 3-kinase (PI3K) which is an activator of AKT and coimmunoprecipitates with PI3K subunits.[10,66] Furthermore, in cell culture model systems, overexpression of GRP78 leads to increased production of phosphatidylinositol-(3, 4, 5)-triphosphate (PIP3) and mutation of the NH_2-terminal region of GRP78 reduces both the binding of CS-GRP78 to PI3K and PIP3 production.[10] A requirement for GRP78 during serum stimulated increase in PIP3 production has also been reported in human leukemic cells.[67] Although, GRP78 is an upstream regulator of the PI3K/AKT pathway, GRP78 is also a downstream target of AKT—an important signal transduction pathway in endometrial cancer.[68] Through interaction with α_2M^*, CS-GRP78 promotes 1-LN prostate cancer cell survival by activating AKT and nuclear factor-κB (NF-κB) signaling cascades.[69] However, it is also possible that as a major molecular chaperone in the ER, GRP78 may be required for processing important growth factors and the cell surface expression of their corresponding receptors which might regulate AKT activation.

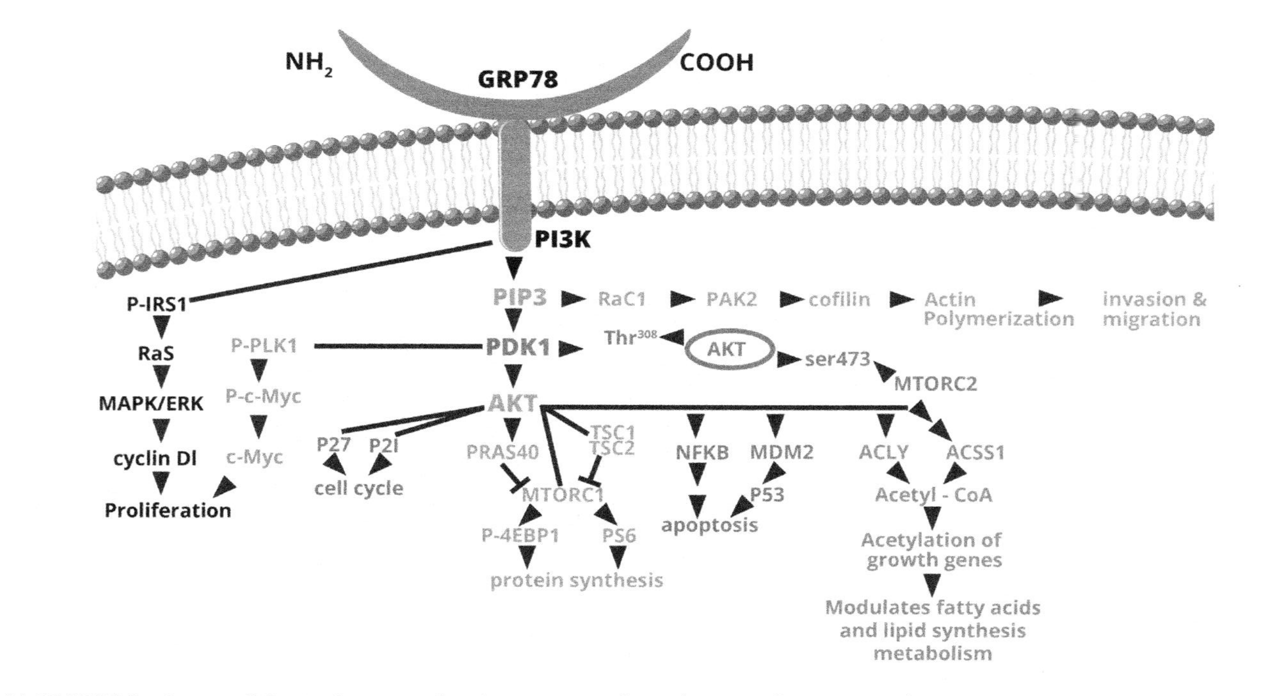

Figure 2.3 CS-GRP78 functions as multifunctional receptor and regulates various signaling pathways to induce tumor growth.

CS-GRP78, through its interaction with different ligands and cell surface proteins, can mediate important signal transduction pathways and is consistent with its novel role as a regulator of the PI3K/AKT signaling pathway—which promotes cell proliferation, survival, metastasis, and chemoresistance.[2,4] α_2M*/CS-GRP78 signaling upregulates mTORC1 and mTORC2 activation and promotes protein synthesis in prostate cancer cells.[70] α_2M*/CS-GRP78 signaling is required for mechanistic target of rapamycin (mTOR) complex-mediated phosphorylation of AKT by 3-phosphoinositide-dependent protein kinase-1 (PDK1).[71] α_2M*/CS-GRP78 acts as an upstream regulator of the PDK1/PLK1 signaling axis to modulate transcriptional activation of proliferative genes.[72] In prostate cancer cell lines, α_2M*/CS-GRP78 signaling regulates PDK1 to phosphorylates AKT in the T-loop at (T308) by using Raptor in the mTORC1 complex as a scaffold protein.[71] Binding of α_2M* to the CS-GRP78 causes its autophosphorylation[73] triggering downstream signaling cascades, which include RAS/MAPK, and PI 3-kinase/Akt/mTOR to induce cellular proliferation and promotes survival.[41,55,69,74–76] As a result, the TSC1−TSC2 complex is phosphorylated inhibiting its GAP activity toward Rheb, thus freeing it for mTORC1 activation. Likewise, phosphorylation of PRAS40 and Deptor by AKT causes their dissociation from mTOR and promotes mTORC1 activation. α_2M* induced phosphorylation of PIP2 by PI3K promotes membrane recruitment of PDK1 and AKT1 where PDK1 phosphorylates AKT at Thr^{308} and mTORC2 phosphorylates AKT at Ser^{473} causing its full activation.[70] Therefore, CS-GRP78 is a potent regulator of mTOR signaling pathway to induce cancer cell proliferation and survival.

In macrophages, α_2M*/CS-GRP78 signaling promotes mitogenic signaling by activating Ras GTP to induce the MEK-ERK pathway.[74,77] The α_2M*/CS-GRP78 axis also induces cell proliferation in prostate cancer by activating antiapoptotic signaling through ERK1/2 and p38 MAPK pathway in addition to its role in Akt activation.[69] CS-GRP78 is also required for Cripto-dependent activation of the MAPK pathway and mitogenic effects in tumor.[64] In Rheumatoid arthritis (RA) patients over expressed cit-GRP78 CS/GRP78 on the surface of rheumatoid synovial cells selectively activates ERK1/2 and JNK signaling pathways to enhance IKK-α phosphorylation which leads to the activation of NF-KB and the production of TNF-α.[78]

α_2M^*/CS-GRP78 regulates the tumor suppress protein P53 through its phosphorylated and acetylation sites. CS-GRP78 ligation by α_2M^* induces activation of PAK2 (p21-activated kinase-2) and, together with LIMK1 and cofilin phosphorylation, increases motility for metastasis.[75,76] Moreover, in human breast cancer, CS-GRP78 regulates the STAT3 pathway in an indirect manner.[79] Secreted GRP78 interacts with the EGFR receptor tyrosine kinase to activate EGFR-SRC-STAT3 signaling which promotes the proliferation of HCC cells and confers the resistance to sorafenib in an autocrine or paracrine manner.[80]

CS-GRP78/Cripto complex alters transforming growth factor β-signaling either directly, by forming complexes with TGF-β ligands and their signaling receptors, or indirectly by activating Src/MAPK/PI3K and possibly Notch and Wnt pathways—which can then engage in crosstalk with the SMAD2/3 pathway. Therefore, CS-GRP78/Cripto complexes act via these direct and indirect mechanisms to cause a switch in Smad2/3 signaling from cytostatic to oncogenic signaling to enhance cell growth.[7] Additionally, CS-GRP78/Cripto signaling has a special role as a gatekeeper for HSCs under hypoxia by regulating the HIF-1 complex.[81] Therefore targeting CS-GRP78 with monoclonal antibodies directed against the COOH-terminal domain of GRP78 could be highly effective in suppressing important signal transduction pathways including AKT activation in a variety of tumors. We have developed several such antibodies including C38 and C107.

BIOLOGICAL FUNCTIONS OF CS-GRP78 IN CANCER

Cell Survival and Proliferation

ER stress promotes the localization of GRP78 on the cell surface of cancer cells which is critical for cell proliferation. Previously, we identified GRP78 primary amino acid sequence LIGRTWNDPSVQQDOKFL (Leu^{98}–Leu^{115}) as the putative binding site for α_2M^*—which is essential for signaling ERK and AKT activation, as well as increased DNA and protein synthesis leading to proliferation and survival.[21,70,72] Autoantibodies against the NH_2-terminal domain of GRP78 are found in the sera of cancer patients. These antibodies mimic α_2M^* to induce proliferation of cancer cells by regulating AKT signaling—which both promotes proliferation and inhibits apoptosis.[82] The antibodies against the NH_2-terminal domain of GRP78 function similarly to α_2M^* to induce proliferation, whereas antibodies against the COOH-terminal domain block apoptotic signaling by the P53 pathway resulting in growth inhibition and cell death.[32] CS-GRP78 regulates IGF-1R activation to promote hepatoma cell proliferation.[61] Additionally, CS-GRP78/Cripto complexes potentiate proliferative signaling by regulating TGFβ signaling in human cancer.[64] Another pro-proliferative mechanism of GRP78 is that α_2M^*/CS-GRP78-dependent PDK1/PLK1 signaling is required for the transcriptional activation of a subset of *c-Myc* target genes and cell proliferation.[72]

APOPTOSIS

α_2M^*/CS-GRP78 complex promotes cancer cell survival by activating the AKT and NF-κB signaling cascades.[69] In hypoxia, CS-GRP78 functions as a receptor for Kringle 5 and on internalization the complex, competes for procaspase 7 binding to the ATP binding domain of GRP78 in the ER which leads to caspase 7 activation and tumor cell apoptosis.[43] Binding of the amino terminal domain of GRP78 with either Kringle 5 or secreted Par-4 elicits cells death. CS-GRP78 binding with PAR4 elicits the extrinsic apoptotic pathway by activation of caspases-3 and 8, whereas CS-GRP78/Kringle 5 mediated apoptosis involves the intrinsic pathway by activation of caspase 7.[17,28,43,83] Thus, the effect of CS-GRP78 on TRAIL-induced apoptosis might be context dependent. The antibodies against the COOH-terminal domain, C38, and C107 induce apoptosis both in vivo and in vitro.[32,82]

INVASION AND MIGRATION

Tumor metastasis is a multistep process that involves the degradation of the ECM, tumor cell migration and invasion, the induction of angiogenesis, as well as tumor cell survival in new tissues. The level of CS-GRP78 is increased in metastatic cancer cell lines, lymph node metastases and human metastatic lesions, and is clearly implicated in the development of metastatic prostate cancer.[84–86] Moreover, CS-GRP78 functioning as a coreceptor for ligands that signal the activation of kinases known to enhance migration such as AKT, FAK, and p21-activated kinase 2 (PAK2).[57,76] It has also been proposed that CS-GRP78 functions as a bridge protein for the uPA-uPAR protease system—which can mediate degradation of the ECM and promote invasion.[57] In particular, peptides targeting α_2M^* and CS-GRP78 are clearly implicated in the development of metastatic androgen-independent prostate cancer.[87] Additionally, CS-GRP78/Cripto signaling regulates the invasive program of metastatic prostate cancer cells that maintains the stem cell-like and aggressive phenotype in human prostate cancer.[65] In hepatoma cells, CS-GRP78 interacts with IGF-1R and induces its phosphorylation and activation to promote hepatoma cell migration and tumor growth.[61] Furthermore, secreted GRP78 activates STAT3 pathway through PI3K/AKT signaling to promote breast cancer cell growth and migration.[79]

TRANSCRIPTIONAL ACTIVATION

α_2M^* stimulation induces the transcriptional activation of TFII-1 which binds to the GRP78 promoter thereby enhancing transcriptional upregulation of GRP78—which promotes tumor cell proliferation and tumor growth (Fig. 2.4). α_2M^* induced TFII-1 also regulates c-Fos transcriptional activation to promote cell proliferation.[29] Additionally, CS-GRP78 regulates the activation of Smad2/3 and STAT3

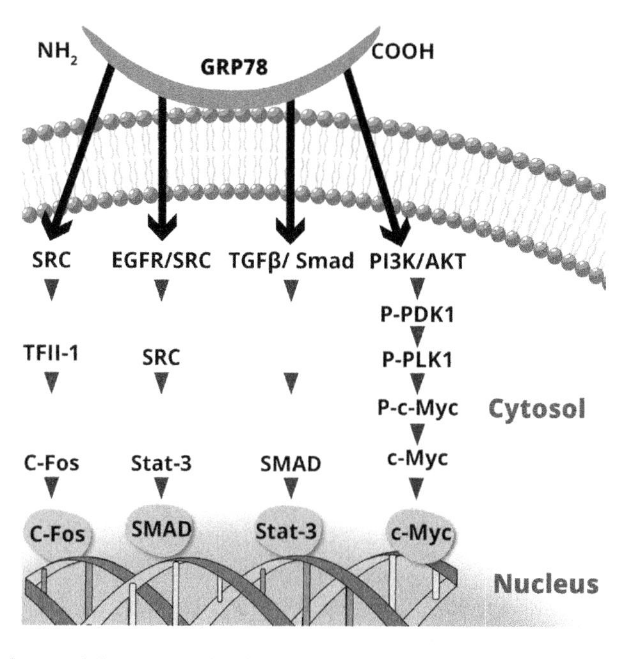

Figure 2.4 Mechanism of CS-GRP78-mediated transcriptional activation of target genes. CS-GRP78 regulates transcription factors like TFII-1, c-Fos, c-MYC, Stat-3, and SMAD. Moreover, CS-GRP78 modulates c-MYC transcription factors to regulate more than 30 c-MYC target genes for proliferation and cancer cell survival.

transcription factors which causes a switch from cytostatic to oncogenic pathways.[64,79,80] The oncogene *c-Myc* globally reprograms cells and drives proliferation by regulating an estimated 15% of the genes in the human genome.[88] Moreover, rather than acting as a general amplifier of transcription, *c-Myc* activates and represses transcription of discrete gene sets leading to change in cell proliferation, tumor progression, and maintenance.[89] We have demonstrated a new role for the α_2M^*/CS-GRP78 signaling axis as a *c-Myc* dependent modifier of chromatin. Moreover, α_2M^*/CS-GRP78-dependent PDK1/PLK1 signaling is required for the transcriptional activation of a subset of *c-Myc* target genes and cell proliferation.[72] These studies extend the role of α_2M^*/CS-GRP78 beyond signal transduction pathways into a transcription activator of target genes.

METABOLISM

Metabolism is an essential mechanism in cancer cell biology. Cancer cells modulate metabolic adaptation for their survival under hypoxia and various other stresses. α_2M^* treatment of prostate cancer cells enhances the Warburg effect and upregulates lipid metabolism in an insulin-like manner. α_2M^*/CS-GRP78 signaling in addition to its ability to induce aerobic glycolysis also functions as a regulator of fatty acid synthesis to promote cancer cell proliferation.[90] We have demonstrated that α_2M^*/CS-GRP78 regulates acetate to function as an epigenetic metabolite to promote cancer cell survival under hypoxic stress. Moreover, α_2M^*/CS-GRP78 signaling axis regulates ATP-citrate lyase (ACLY) and Acetyl-CoA synthetase (ACSS1) expression by promoting AKT activation, this ensures continuous acetyl-CoA production and global histone acetylation even during nutrient limitation.[91] CS-GRP78 regulates complex mechanisms in cancer cells in that it modulates metabolic regulation which might be an adaptive mechanism in rapidly proliferating tumors.

THE ROLE OF CS-GRP78 IN OTHER PHYSIOLOGICAL DISEASES

Angiogenesis

Eliminating the tumor vasculature—which supplies nutrients and oxygen within the tumor—is a key strategy for anticancer therapy. CS-GRP78 is highly expressed on the proliferating endothelial cells.[43,52] VEGF can induce CS-GRP78 in endothelial cells, while knockdown of

GRP78 suppresses VEGF-mediated MAPK signaling and endothelial cell proliferation.[36] CS-GRP78 on the endothelial cells also induces VEGF independent angiogenesis under hypoxic conditions, implying it as an angiogenic receptor for ischemic disease therapy.[92] The cell surface voltage dependent anion channel (VDAC) interacts with CS-GRP78 to induce the apoptotic effect of Kringle 5 in proliferating endothelial cells.[28] CS-GRP78/T-Cadherin complex regulates T-Cadherin-dependent endothelial cell survival.[62] CS-GRP78 also interacts with Kringle 5 in glioma endothelial cells to induce apoptosis[43] that can be increased by radiation, as this results in the internalization of CS-GRP78 by LRP1 and in the activation of p38 MAPK.[56] Collectively, these studies suggest that targeting CS-GRP78 could achieve a dual effect in suppressing tumor growth as well as in tumor angiogenesis.

VIRUS INTERNALIZATION

GRP78 acts as a coreceptor for virus internalization by associating with MHC class I molecules on the cell surface.[93] Viral entry of Coxsackievirus A9 into host cells required CS-GRP78-mediated MHC class I molecules. GRP78 expressed on the surface of liver cancer cells acts as a receptor for dengue virus serotype 2 entry and antibodies directed against both the NH_2- and COOH-terminal domains of GRP78 had a major effect on the binding of the virus to the cell surface as well as virus infectivity.[94] Borna disease virus entry was mediated by the association of CS-GRP78 with the NH_2-terminus cleaved product of the envelope glycoprotein of Borna disease virus, GPI.[95] An antibody against the NH_2-terminal domain of GRP78 (N20) inhibits GPI binding to cells expressing GRP78 on the cell surface and to reduce virus infection.

INFLAMMATION AND IMMUNITY

MHC-I downregulation is a widespread mechanism used by tumor cells to evade immune surveillance.[51,96,97] GRP78 is an obligatory binding partner for cell surface MHC class I molecules.[58] The downregulation of MHC-1 expression in cells overexpressing GRP78 not only protects the tumor cell from immune surveillance, but also promotes tumor progression. α_2M^*/CS-GRP78 signaling regulates the G_s-mediated cyclic AMP (cAMP) production and the proinflammatory cyclooxygenase 2 (COX2)-prostaglandin E (PGE)-cAMP signaling cascade.[98,99] CS-GRP78 in T cells forms a complex with, and confers stabilization to, cell surface TGFβ—which are an immune regulator and an inducer of T_{Reg} cells.[100] Some cancer cells secrete GRP78, which modulates the differentiation of human monocytes into mature dendritic cells, while subsequent recruitment of T cells leads to the generation of T_{Reg} cells.[101] Interaction of MTJ-1 with GRP78 is essential for transport of GRP78 into the cell surface where it serves a number of functions in immune regulation and signal transduction.[16] In addition, α_2M^*/CS-GRP78 complexes modulate immune responses as well as promote macrophage locomotion and chemotaxis.[73] Thus CS-GRP78 regulates inflammation and immunity both in tumor cells and through interactions with the tumor microenvironment.

ARTHRITIS

GRP78 expresses on the cell surface and acts as an autoantigen in RA. Indeed, 30%–75% of RA patients exhibit CS-GRP78 in synovial fluid or GRP78 autoantibodies in serum.[102–104] Although the exact roles of CS-GRP78 and GRP78 autoantibodies in RA pathogenesis remain debatable, three possible functions can be considered. First, CS-GRP78 and its antibody may result from inflammatory stress during RA progression and might have no pathophysiological roles. The unfolded protein response is stimulated through inflammation and ER stress proteins, including GRP78, are subsequently induced,[105] GRP78 could be simply a specific and abundant autoantigen in the inflammatory lesion that is unrelated to RA pathogenesis. Second, the autoantigen might act as an antiinflammatory factor, similar to HSPs. Several studies suggest that CS-GRP78 stimulates the production of the antiinflammatory cytokines, IL-4, and IL-10, through specific T lymphocytes.[102,103,106–109] Furthermore, the preadministration of GRP78 protein to mice prevents the in vivo induction of adjuvant arthritis or CIA, both of which are well-known artificial models of autoimmune diseases.[103,108] Third, the CS-GRP78 antigen could act as a proinflammatory factor. GRP78 positively causes immunological responses and inflammation. Recently, Shoda et al.[104], reported that in addition to the intact GRP78 antibody, the anticitrullinated GRP78 (CITGRP78) antibody is frequently detected in RA patients. The CITGRP78 protein, but not intact GRP78, enhances anticitrullin antibodies and worsen arthritis symptoms in mouse models of adjuvant arthritis. Citrullination is a protein modification in which an arginine

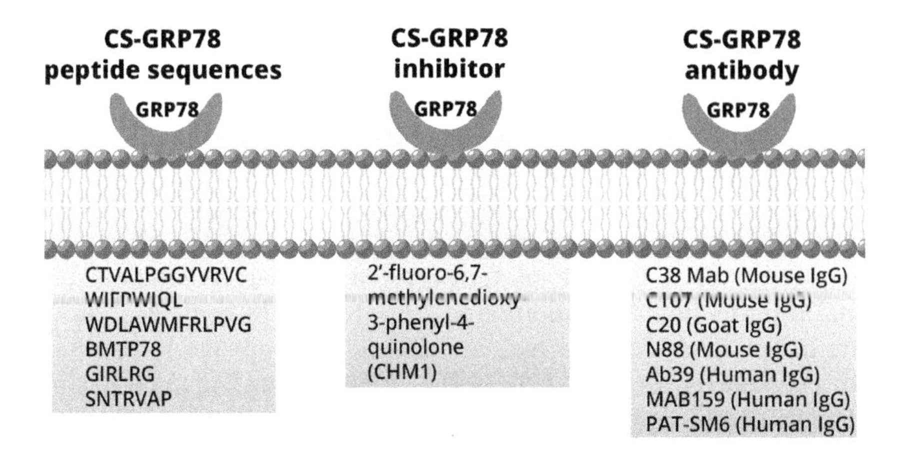

Figure 2.5 Three classes of CS-GRP78 have been developed for cancer therapy. Anti-CS-GRP78 antibody, peptide sequences that block the CS-GRP78 signaling activity, and small molecule inhibitors are in preclinical and clinical development for the treatment of cancer.

residue is converted into a citrulline residue by a specific intracellular enzyme—peptidylarginine deiminase (Pad).[110] Anticitrullinated peptide/protein antibodies are frequently detected in RA patients.[111–113] Although the reasons why citrullination is frequently observed and how it participates in RA pathogenesis remains unclear, the relationship between stress proteins and this specific protein modification suggests an undescribed crosstalk between inflammatory stress and disease specific protein modifications in RA pathogenesis. In addition to RA, the GRP78 autoantibody is also detected in another autoimmune and inflammatory disease, systemic lupus erythematosus[114,115] in which its pathophysiological role remains unknown.

THERAPEUTIC TARGETING OF CS-GRP78

Clinical and preclinical studies indicate an important role for CS-GRP78 in tumor progression, metastasis, and drug resistance. As CS-GRP78 is preferentially expressed on tumor cells in vivo, it may serve as a specific tumor target with minimal harmful effects on normal cells.[4,6,14] Moreover, CS-GRP78 is detected in advanced clinical conditions and is increased in metastatic, drug resistant tumors.[10,116,117] Radiation also upregulates CS-GRP78. Moreover, hypoxic endothelial cells that support tumor growth, show upregulation of CS-GRP78. Thus, therapeutic agents against cell surface GRP78 have the potential to target these cells in addition to the primary tumor (Fig. 2.5).

Several synthetic GRP78 binding peptides are able to promote apoptosis in cancer cells in vitro, including human prostate cancer cells, human breast cancer cells, human melanoma, chemotherapy resistant B-lineage acute lymphoblastic leukemia cells, and multidrug-resistant gastric cells.[50,51,72,116,118–122] Furthermore, xenograft and isogenic mouse models were used to validate the efficacy of the peptides in suppressing the growth of prostate cancer, breast cancer, melanoma, and bone metastasis with no apparent toxicity.[87,116,118,119] For more efficient drug delivery the GRP78 binding peptides have been conjugated to nanoparticles or liposomes which are able to enter into the endothelial cells in tumors,[119,123] thereby suppressing tumor growth in colon cancer and prolonging their survival.[36] The GRP78 binding peptide CTVALPGGYVRVC interacts at the COOH-terminal region to abrogate signaling and c-Myc transcriptional activity.[21,72] Whereas the other GRP78 binding peptides, WIFPWIQL and WDLAWMFRLPVG, induce apoptosis to abrogate tumor growth.[118] By using a peptidomimetic targeting strategy it has been shown that the bone metastasis targeting peptide 78 (BMTP78) selectively kills breast cancer cells that express surface localized GRP78.[116] Furthermore, in preclinical metastasis models, administration of BMTP78 can inhibit primary tumor growth as well as prolong overall survival by reducing the extent of outgrowth of established lung and bone micro metastases. GIRLRG peptide specifically binds to the CS-GRP78 which functions as a peptide receptor and a molecular target of diagnostics and therapeutics of various cancer.[124] Recently CS-GRP78 has been targeted by engineered human Herpes simplex virus thymidine kinase type-1 as a noninvasive imaging reporter/suicide transgene into adeno-associated virus phage (AAVP) particles displaying motif ligands SNTRVAP to CS-GRP78 in aggressive human breast and prostate cancer.[125,126] The AAVP-based strategy has maximum diagnostic and therapeutic efficacy targeted at the tumor site, by simultaneously image and monitors the aggressive tumors which making CS-GRP78 as a promising candidate for the development of human clinical trials. These studies suggest that different GRP78 targeting peptides from different tumors validate GRP78 as an ideal tumor target for clinical intervention.

CHM-1 is the only chemical inhibitor that suppresses the GRP78/PI3K/AKT signaling complex by inhibiting the formation of cell surface-associated GRP78-p85α complexes to promote apoptosis of human nasopharyngeal carcinoma cells.[127] The NH_2-terminal domain-reactive antibody, mimics α_2M^* as a ligand and drives PI 3-kinase-dependent activation of AKT and the subsequent stimulation of cell proliferation in vitro.[128] Monoclonal antibodies (Mab) developed by our laboratory, C38 and C107, and a commercial antibody (C20) directed against the COOH-terminal domain of GRP78, suppress tumor growth in prostate and melanoma models.[32,128] The COOH-terminal domain reactive antibody C38 acts as an antagonist of both α_2M^* and N88Mab by activating a caspase pathway resulting in delayed tumor growth and prolonged survival in a mouse melanoma and ovarian model.[39,128] N88Mab binds to the NH_2-terminal domain of GRP78 and mimics human autoantibodies. Moreover, C38Mab targets cancer stem-like cells and prolongs survival of ovarian cancer-bearing mice.[39] Mouse monoclonal immunoglobulin G (IgG) against the amino acid sequence KDEL, in the COOH-terminal region of anti-GRP78, inhibits cellular proliferation and induces apoptosis.[129] Conversely, a human monoclonal GRP78 antibody against a region in the last COOH-terminal 20 amino acid residues does not affect cell proliferation or induce apoptosis.[14] Another screen yielded a mouse monoclonal IgG antibody that targets the COOH-terminal domain of GRP78, C107, induces apoptosis in melanoma cells in vitro, and slows their growth in mice.[128]

Human antibody Ab39 recognizes COOH-terminal region of CS-GRP78 and suppresses the proliferation and tumor growth of varies tumors. This makes Ab39 an attractive therapeutic agent to suppress CS-GRP78 on cancer cells.[14] Mab159, a GRP78-specific mouse monoclonal IgG antibody causes cancer cell death and suppresses the growth of colon and lung xenografts, the metastatic growth of human breast and human melanoma xenografts, and the growth of prostate cancer and leukemia in genetically engineered mouse models—at least in part through inhibition of the PI3K signaling pathway.[66] Moreover, Mab159 was humanized and validated for diagnostic and therapeutic application. A human monoclonal IgM antibody, PAT-SM6, which was isolated from a patient with gastric cancer which induces apoptosis in multiple myeloma cells and suppresses human melanoma growth both in vitro and in xenografts.[130] Based on favorable safety profiles in Phase I studies, the efficacy of PAT-SM6 is being tested in clinical trials.[131] Moreover, in human trials PAT-SM6 therapy in combination with bortezomib and lenalidomide showed efficacy in relapsed refractory multiple myeloma.[132] Thus, significant data suggest that CS-GRP78 may be amenable as a therapeutic target in the suppression of tumor growth and translational applications in aggressive tumor variants.

GRP78 AUTOANTIBODIES AND THEIR PHYSIOLOGICAL ROLES

The GRP78 autoantibody isolated from prostate cancer patients binds to Leu^{98}–Leu^{115}, which corresponds to the α_2M^* binding sites.[21] Similar to its association with α_2M^*, binding of GRP78 to the GRP78 autoantibody promotes cell survival and proliferation. In contrast, COOH-terminal anti-GRP78 antibodies act as receptor antagonists by blocking auto phosphorylation and activation of GRP78.[32] Autoantibodies against GRP78 from patients with ovarian cancer promote apoptosis and decrease the invasiveness of ovarian cancer cells.[133] In another study, autoantibodies against GRP78 from patients with prostate cancer triggered ER Ca^{2+} release in human bladder carcinoma cells and increased tissue factor PCA—which implies that blocking CS-GRP78 signaling could potentially reduce the risk of cancer-related thrombotic events.[54] GRP78 may be a putative autoantigen in RA since there is a presence of GRP78 autoantibodies in the synovial fluid of these patients, while there are no detectable antibodies in healthy subjects.[102,103]

CONCLUDING REMARKS

CS-GRP78 has emerged as a promising anticancer target. Notably, the majority of tumors induce CS-GRP78 during tumor progression, metastasis, and resistance. Targeting CS-GRP78 signaling in tumor models has begun to illuminate the diverse mechanisms by which the signaling pathways promote tumor progression and metastasis. However, important questions still remain in the field.

Tumor stromal crosstalk plays an important role in tumor initiation, progression, and response to therapy. Tumor cells communicate with a variety of stromal cells including fibroblasts, bone marrow-derived cells, immune vascular and epithelial cells to facilitate tumor progression, metastasis, and therapy resistance.[134] The majority of work in the field has focused on the role of CS-GRP78 signaling in tumor cells and its impact on tumor behavior. With the use of C38

Mab in immune competent tumor models, we will be able to better understand how CS-GRP78 signaling is utilized in tumor stromal crosstalk to promote tumor progression. Studies have indicated that CS-GRP78 signaling has the potential to influence the tumor and stromal cells in prostate cancer.[87] Future studies investigating the role of CS-GRP78 signaling within individual stromal cell populations are needed to better understand the diverse roles of CS-GRP78 signaling within the tumor microenvironment. In addition, it will be important to further elucidate the role of CS-GRP78 signaling in mediating receptor tyrosine kinase crosstalk. The TAM family members have been shown to crosstalk and cooperate with each other and with other RTKs in the activation of downstream signaling events.[135] These studies may have important implications in therapeutic resistance and indicate novel combination therapies for the treatment of cancer.

The basis of many human diseases arises from both genetic and epigenetic regulation. Advances in the understanding of the mechanisms underlying transcriptional and epigenetic regulation and their prevalence as contributors to a cancer's progression have led us to focus on transcription and epigenetic changes in cancer. Our studies demonstrate that α_2M^*/CS-GRP78 acts as an upstream regulator of the PDK1/PLK1 signaling axis to modulate *c-MYC* transcription and its target genes, suggesting that CS-GRP78 might be potential therapeutic target in *c-MYC* driven tumors.[72] Although our current studies focus on characterizing the role of CS-GRP78 in the transcriptional factor activation, we have extended our investigation to include epigenetic mechanisms as major determinants of gene activation, in particular histone acetylation. Therefore, it is essential to focus on studying transcriptional and epigenetic changes to understand the fundamental mechanisms of how CS-GRP78 plays a critical role in the regulation of gene expression in cancer cells.

Therapeutic inhibition of CS-GRP78 signaling has been utilized to identify important functional roles for CS-GRP78 signaling in tumor progression, metastasis, and drug resistance. As a result, there are varieties of Mab and peptides targeting CS-GRP78 available for the diagnosis and treatment of cancer. Future studies are needed to thoroughly investigate the cellular and molecular mechanisms by which CS-GRP78 signaling promotes tumor progression in order to develop the most effective anti-CS-GRP78 combination therapies.

REFERENCES

1. Ni M, Lee AS. ER chaperones in mammalian development and human diseases. *FEBS Lett* 2007;**581**:3641–51.
2. Lee AS. Glucose-regulated proteins in cancer: molecular mechanisms and therapeutic potential. *Nat Rev Cancer* 2014;**14**:263–76.
3. Wang M, Wey S, Zhang Y, Ye R, Lee AS. Role of the unfolded protein response regulator GRP78/BiP in development, cancer, and neurological disorders. *Antioxid Redox Signal* 2009;**11**:2307–16.
4. Ni M, Zhang Y, Lee AS. Beyond the endoplasmic reticulum: atypical GRP78 in cell viability, signalling and therapeutic targeting. *Biochem J* 2011;**434**:181–8.
5. Gonzalez-Gronow M, Selim MA, Papalas J, Pizzo SV. GRP78: a multifunctional receptor on the cell surface. *Antioxid Redox Signal* 2009;**11**:2299–306.
6. Sato M, Yao VJ, Arap W, Pasqualini R. GRP78 signaling hub a receptor for targeted tumor therapy. *Adv Genet* 2010;**69**:97–114.
7. Gray PC, Vale W. Cripto/GRP78 modulation of the TGF-beta pathway in development and oncogenesis. *FEBS Lett* 2012;**586**:1836–45.
8. Berger CL, Dong Z, Hanlon D, Bisaccia E, Edelson RL. A lymphocyte cell surface heat shock protein homologous to the endoplasmic reticulum chaperone, immunoglobulin heavy chain binding protein BIP. *Int J Cancer* 1997;**71**:1077–85.
9. Zhang Y, Liu R, Ni M, Gill P, Lee AS. Cell surface relocalization of the endoplasmic reticulum chaperone and unfolded protein response regulator GRP78/BiP. *J Biol Chem* 2010;**285**:15065–75.
10. Zhang Y, Tseng CC, Tsai YL, Fu X, Schiff R, Lee AS. Cancer cells resistant to therapy promote cell surface relocalization of GRP78 which complexes with PI3K and enhances PI(3,4,5)P3 production. *PLoS One* 2013;**8**:e80071.
11. Shin BK, Wang H, Yim AM, Le Naour F, Brichory F, Jang JH, et al. Global profiling of the cell surface proteome of cancer cells uncovers an abundance of proteins with chaperone function. *J Biol Chem* 2003;**278**:7607–16.
12. Munro S, Pelham HR. A C-terminal signal prevents secretion of luminal ER proteins. *Cell* 1987;**48**:899–907.
13. Llewellyn DH, Roderick HL, Rose S. KDEL receptor expression is not coordinatedly up-regulated with ER stress-induced reticuloplasmin expression in HeLa cells. *Biochem Biophys Res Commun* 1997;**240**:36–40.
14. Jakobsen CG, Rasmussen N, Laenkholm AV, Ditzel HJ. Phage display derived human monoclonal antibodies isolated by binding to the surface of live primary breast cancer cells recognize GRP78. *Cancer Res* 2007;**67**:9507–17.
15. Rauschert N, Brandlein S, Holzinger E, Hensel F, Muller-Hermelink HK, Vollmers HP. A new tumor-specific variant of GRP78 as target for antibody-based therapy. *Lab Invest* 2008;**88**:375–86.
16. Misra UK, Gonzalez-Gronow M, Gawdi G, Pizzo SV. The role of MTJ-1 in cell surface translocation of GRP78, a receptor for alpha 2-macroglobulin-dependent signaling. *J Immunol* 2005;**174**:2092–7.
17. Burikhanov R, Zhao Y, Goswami A, Qiu S, Schwarze SR, Rangnekar VM. The tumor suppressor Par-4 activates an extrinsic pathway for apoptosis. *Cell* 2009;**138**:377–88.
18. Takemoto H, Yoshimori T, Yamamoto A, Miyata Y, Yahara I, Inoue K, et al. Heavy chain binding protein (BiP/GRP78) and endoplasmin are exported from the endoplasmic reticulum

in rat exocrine pancreatic cells, similar to protein disulfide-isomerase. *Arch Biochem Biophys* 1992;**296**:129–36.

19. Kern J, Untergasser G, Zenzmaier C, Sarg B, Gastl G, Gunsilius E, et al. GRP-78 secreted by tumor cells blocks the antiangiogenic activity of bortezomib. *Blood* 2009;**114**:3960–7.
20. Tsunemi S, Nakanishi T, Fujita Y, Bouras G, Miyamoto Y, Miyamoto A, et al. Proteomics-based identification of a tumor-associated antigen and its corresponding autoantibody in gastric cancer. *Oncol Rep* 2010;**23**:949–56.
21. Gonzalez-Gronow M, Cuchacovich M, Llanos C, Urzua C, Gawdi G, Pizzo SV. Prostate cancer cell proliferation in vitro is modulated by antibodies against glucose-regulated protein 78 isolated from patient serum. *Cancer Res* 2006;**66**:11424–31.
22. Marin-Briggiler CI, Gonzalez-Echeverria MF, Munuce MJ, Ghersevich S, Caille AM, Hellman U, et al. Glucose-regulated protein 78 (Grp78/BiP) is secreted by human oviduct epithelial cells and the recombinant protein modulates sperm-zona pellucida binding. *Fertil Steril* 2010;**93**:1574–84.
23. Record M, Subra C, Silvente-Poirot S, Poirot M. Exosomes as intercellular signalosomes and pharmacological effectors. *Biochem Pharmacol* 2011;**81**:1171–82.
24. Skog J, Wurdinger T, van Rijn S, Meijer DH, Gainche L, Sena-Esteves M, et al. Glioblastoma microvesicles transport RNA and proteins that promote tumour growth and provide diagnostic biomarkers. *Nat Cell Biol* 2008;**10**:1470–6.
25. Taylor DD, Gercel-Taylor C, Parker LP. Patient-derived tumor-reactive antibodies as diagnostic markers for ovarian cancer. *Gynecol Oncol* 2009;**115**:112–20.
26. Xiao D, Ohlendorf J, Chen Y, Taylor DD, Rai SN, Waigel S, et al. Identifying mRNA, microRNA and protein profiles of melanoma exosomes. *PLoS One* 2012;**7**:e46874.
27. Li Z, Zhuang M, Zhang L, Zheng X, Yang P, Li Z. Acetylation modification regulates GRP78 secretion in colon cancer cells. *Sci Rep* 2016;**6**:30406.
28. Gonzalez-Gronow M, Kaczowka SJ, Payne S, Wang F, Gawdi G, Pizzo SV. Plasminogen structural domains exhibit different functions when associated with cell surface GRP78 or the voltage-dependent anion channel. *J Biol Chem* 2007;**282**:32811–20.
29. Misra UK, Wang F, Pizzo SV. Transcription factor TFII-I causes transcriptional upregulation of GRP78 synthesis in prostate cancer cells. *J Cell Biochem* 2009;**106**:381–9.
30. Chen LY, Chiang AS, Hung JJ, Hung HI, Lai YK. Thapsigargin-induced GRP78 expression is mediated by the increase of cytosolic free calcium in 9L rat brain tumor cells. *J Cell Biochem* 2000;**78**:404–16.
31. Miyake H, Hara I, Arakawa S, Kamidono S. Stress protein GRP78 prevents apoptosis induced by calcium ionophore, ionomycin, but not by glycosylation inhibitor, tunicamycin, in human prostate cancer cells. *J Cell Biochem* 2000;**77**:396–408.
32. Misra UK, Mowery Y, Kaczowka S, Pizzo SV. Ligation of cancer cell surface GRP78 with antibodies directed against its COOH-terminal domain up-regulates p53 activity and promotes apoptosis. *Mol Cancer Ther* 2009;**8**:1350–62.
33. Thastrup O, Dawson AP, Scharff O, Foder B, Cullen PJ, Drobak BK, et al. Thapsigargin, a novel molecular probe for studying intracellular calcium release and storage. 1989. *Agents Actions* 1994;**43**:187–93.
34. Treiman M, Caspersen C, Christensen SB. A tool coming of age: thapsigargin as an inhibitor of sarco-endoplasmic reticulum Ca(2 +)-ATPases. *Trends Pharmacol Sci* 1998;**19**:131–5.
35. Yoshida I, Monji A, Tashiro K, Nakamura K, Inoue R, Kanba S. Depletion of intracellular Ca2+ store itself may be a major factor in thapsigargin-induced ER stress and apoptosis in PC12 cells. *Neurochem Int* 2006;**48**:696–702.

36. Katanasaka Y, Ishii T, Asai T, Naitou H, Maeda N, Koizumi F, et al. Cancer antineovascular therapy with liposome drug delivery systems targeted to BiP/GRP78. *Int J Cancer* 2010;**127**:2685–98.

37. Bellani S, Mescola A, Ronzitti G, Tsushima H, Tilve S, Canale C, et al. GRP78 clustering at the cell surface of neurons transduces the action of exogenous alpha-synuclein. *Cell Death Differ* 2014;**21**:1971–83.

38. Raiter A, Yerushalmi R, Hardy B. Pharmacological induction of cell surface GRP78 contributes to apoptosis in triple negative breast cancer cells. *Oncotarget* 2014;**5**:11452–63.

39. Mo L, Bachelder RE, Kennedy M, Chen PH, Chi JT, Berchuck A, et al. Syngeneic murine ovarian cancer model reveals that ascites enriches for ovarian cancer stem-like cells expressing membrane GRP78. *Mol Cancer Ther* 2015;**14**:747–56.

40. Dadey DYA, Kapoor V, Hoye K, Khudanyan A, Collins A, Thotala D, et al. Antibody targeting GRP78 enhances the efficacy of radiation therapy in human glioblastoma and non-small cell lung cancer cell lines and tumor models. *Clin Cancer Res* 2017;**23**:2556–64.

41. Misra UK, Payne S, Pizzo SV. Ligation of prostate cancer cell surface GRP78 activates a proproliferative and antiapoptotic feedback loop: a role for secreted prostate-specific antigen. *J Biol Chem* 2011;**286**:1248–59.

42. Chen M, Zhang Y, Yu VC, Chong YS, Yoshioka T, Ge R. Isthmin targets cell-surface GRP78 and triggers apoptosis via induction of mitochondrial dysfunction. *Cell Death Differ* 2014;**21**:797–810.

43. Davidson DJ, Haskell C, Majest S, Kherzai A, Egan DA, Walter KA, et al. Kringle 5 of human plasminogen induces apoptosis of endothelial and tumor cells through surface-expressed glucose-regulated protein 78. *Cancer Res* 2005;**65**:4663–72.

44. Shrestha-Bhattarai T, Rangnekar VM. Cancer-selective apoptotic effects of extracellular and intracellular Par-4. *Oncogene* 2010;**29**:3873–80.

45. Schwarze S, Rangnekar VM. Targeting plasma membrane GRP78 for cancer growth inhibition. *Cancer Biol Ther* 2010;**9**:153–5.

46. Lee AS. The Par-4-GRP78 TRAIL, more twists and turns. *Cancer Biol Ther* 2009;**8**:2103–5.

47. Nakatsuka A, Wada J, Iseda I, Teshigawara S, Higashio K, Murakami K, et al. Vaspin is an adipokine ameliorating ER stress in obesity as a ligand for cell-surface GRP78/MTJ-1 complex. *Diabetes* 2012;**61**:2823–32.

48. Huergo-Zapico L, Gonzalez-Rodriguez AP, Contesti J, Gonzalez E, Lopez-Soto A, Fernandez-Guizan A, et al. Expression of ERp5 and GRP78 on the membrane of chronic lymphocytic leukemia cells: association with soluble MICA shedding. *Cancer Immunol Immunother* 2012;**61**:1201–10.

49. Dekki N, Refai E, Holmberg R, Kohler M, Jornvall H, Berggren PO, et al. Transthyretin binds to glucose-regulated proteins and is subjected to endocytosis by the pancreatic beta-cell. *Cell Mol Life Sci* 2012;**69**:1733–43.

50. Kim Y, Lillo AM, Steiniger SC, Liu Y, Ballatore C, Anichini A, et al. Targeting heat shock proteins on cancer cells: selection, characterization, and cell-penetrating properties of a peptidic GRP78 ligand. *Biochemistry* 2006;**45**:9434–44.

51. Liu Y, Steiniger SC, Kim Y, Kaufmann GF, Felding-Habermann B, Janda KD. Mechanistic studies of a peptidic GRP78 ligand for cancer cell-specific drug delivery. *Mol Pharm* 2007;**4**:435–47.

52. Bhattacharjee G, Ahamed J, Pedersen B, El-Sheikh A, Mackman N, Ruf W, et al. Regulation of tissue factor--mediated initiation of the coagulation cascade by cell surface grp78. *Arterioscler Thromb Vasc Biol* 2005;**25**:1737–43.

53. Watson LM, Chan AK, Berry LR, Li J, Sood SK, Dickhout JG, et al. Overexpression of the 78-kDa glucose-regulated protein/immunoglobulin-binding protein (GRP78/BiP) inhibits tissue factor procoagulant activity. *J Biol Chem* 2003;**278**:17438–47.

54. Al-Hashimi AA, Caldwell J, Gonzalez-Gronow M, Pizzo SV, Aboumrad D, Pozza L, et al. Binding of anti-GRP78 autoantibodies to cell surface GRP78 increases tissue factor procoagulant activity via the release of calcium from endoplasmic reticulum stores. *J Biol Chem* 2010;**285**:28912–23.

55. Misra UK, Gonzalez-Gronow M, Gawdi G, Hart JP, Johnson CE, Pizzo SV. The role of Grp78 in alpha 2-macroglobulin-induced signal transduction. Evidence from RNA interference that the low density lipoprotein receptor-related protein is associated with, but not necessary for, GRP78-mediated signal transduction. *J Biol Chem* 2002;**277**:42082–7.

56. McFarland BC, Stewart Jr. J, Hamza A, Nordal R, Davidson DJ, Henkin J, et al. Plasminogen kringle 5 induces apoptosis of brain microvessel endothelial cells: sensitization by radiation and requirement for GRP78 and LRP1. *Cancer Res* 2009;**69**:5537–45.

57. Li Z, Zhang L, Zhao Y, Li H, Xiao H, Fu R, et al. Cell-surface GRP78 facilitates colorectal cancer cell migration and invasion. *Int J Biochem Cell Biol* 2013;**45**:987–94.

58. Triantafilou M, Fradelizi D, Triantafilou K. Major histocompatibility class one molecule associates with glucose regulated protein (GRP) 78 on the cell surface. *Hum Immunol* 2001;**62**:764–70.

59. Paulsson KM, Wang P, Anderson PO, Chen S, Pettersson RF, Li S. Distinct differences in association of MHC class I with endoplasmic reticulum proteins in wild-type, and beta 2-microglobulin- and TAP-deficient cell lines. *Int Immunol* 2001;**13**:1063–73.

60. Zhao S, Li H, Wang Q, Su C, Wang G, Song H, et al. The role of c-Src in the invasion and metastasis of hepatocellular carcinoma cells induced by association of cell surface GRP78 with activated alpha2M. *BMC Cancer* 2015;**15**:389.

61. Yin Y, Chen C, Chen J, Zhan R, Zhang Q, Xu X, et al. Cell surface GRP78 facilitates hepatoma cells proliferation and migration by activating IGF-IR. *Cell Signal* 2017;**35**:154–62.

62. Philippova M, Ivanov D, Joshi MB, Kyriakakis E, Rupp K, Afonyushkin T, et al. Identification of proteins associating with glycosylphosphatidylinositol- anchored T-cadherin on the surface of vascular endothelial cells: role for Grp78/BiP in T-cadherin-dependent cell survival. *Mol Cell Biol* 2008;**28**:4004–17.

63. Shani G, Fischer WH, Justice NJ, Kelber JA, Vale W, Gray PC. GRP78 and Cripto form a complex at the cell surface and collaborate to inhibit transforming growth factor beta signaling and enhance cell growth. *Mol Cell Biol* 2008;**28**:666–77.

64. Kelber JA, Panopoulos AD, Shani G, Booker EC, Belmonte JC, Vale WW, et al. Blockade of Cripto binding to cell surface GRP78 inhibits oncogenic Cripto signaling via MAPK/PI3K and Smad2/3 pathways. *Oncogene* 2009;**28**:2324–36.

65. Zoni E, Chen L, Karkampouna S, Granchi Z, Verhoef EI, La Manna F, et al. CRIPTO and its signaling partner GRP78 drive the metastatic phenotype in human osteotropic prostate cancer. *Oncogene* 2017;**36**:4739–49.

66. Liu R, Li X, Gao W, Zhou Y, Wey S, Mitra SK, et al. Monoclonal antibody against cell surface GRP78 as a novel agent in suppressing PI3K/AKT signaling, tumor growth, and metastasis. *Clin Cancer Res* 2013;**19**:6802–11.

67. Wey S, Luo B, Lee AS. Acute inducible ablation of GRP78 reveals its role in hematopoietic stem cell survival, lymphogenesis and regulation of stress signaling. *PLoS One* 2012;**7**: e39047.

68. Gray MJ, Mhawech-Fauceglia P, Yoo E, Yang W, Wu E, Lee AS, et al. AKT inhibition mitigates GRP78 (glucose-regulated protein) expression and contribution to chemoresistance in endometrial cancers. *Int J Cancer* 2013;**133**:21–30.

69. Misra UK, Deedwania R, Pizzo SV. Activation and cross-talk between Akt, NF-kappaB, and unfolded protein response signaling in 1-LN prostate cancer cells consequent to ligation of cell surface-associated GRP78. *J Biol Chem* 2006;**281**:13694–707.

70. Misra UK, Pizzo SV. Receptor-recognized alpha(2)-macroglobulin binds to cell surface-associated GRP78 and activates mTORC1 and mTORC2 signaling in prostate cancer cells. *PLoS One* 2012;**7**:e51735.

71. Misra UK, Pizzo SV. Activated alpha2-macroglobulin binding to cell surface GRP78 induces T-loop phosphorylation of Akt1 by PDK1 in association with Raptor. *PLoS One* 2014;**9**: e88373.

72. Gopal U, Gonzalez-Gronow M, Pizzo SV. Activated alpha2-macroglobulin regulates transcriptional activation of c-MYC target genes through cell surface GRP78 protein. *J Biol Chem* 2016;**291**:10904–15.

73. Misra UK, Sharma T, Pizzo SV. Ligation of cell surface-associated glucose-regulated protein 78 by receptor-recognized forms of alpha 2-macroglobulin: activation of p21-activated protein kinase-2-dependent signaling in murine peritoneal macrophages. *J Immunol* 2005;**175**:2525–33.

74. Misra UK, Pizzo SV. Potentiation of signal transduction mitogenesis and cellular proliferation upon binding of receptor-recognized forms of alpha2-macroglobulin to 1-LN prostate cancer cells. *Cell Signal* 2004;**16**:487–96.

75. Misra UK, Gonzalez-Gronow M, Gawdi G, Wang F, Pizzo SV. A novel receptor function for the heat shock protein Grp78: silencing of Grp78 gene expression attenuates alpha2M*-induced signalling. *Cell Signal* 2004;**16**:929–38.

76. Misra UK, Deedwania R, Pizzo SV. Binding of activated alpha2-macroglobulin to its cell surface receptor GRP78 in 1-LN prostate cancer cells regulates PAK-2-dependent activation of LIMK. *J Biol Chem* 2005;**280**:26278–86.

77. Misra UK, Pizzo SV. Ligation of the alpha2M signalling receptor elevates the levels of p21Ras-GTP in macrophages. *Cell Signal* 1998;**10**:441–5.

78. Lu MC, Lai NS, Yin WY, Yu HC, Huang HB, Tung CH, et al. Anti-citrullinated protein antibodies activated ERK1/2 and JNK mitogen-activated protein kinases via binding to surface-expressed citrullinated GRP78 on mononuclear cells. *J Clin Immunol* 2013;**33**:558–66.

79. Yao X, Liu H, Zhang X, Zhang L, Li X, Wang C, et al. Cell surface GRP78 accelerated breast cancer cell proliferation and migration by activating STAT3. *PLoS One* 2015;**10**: e0125634.

80. Li R, Yanjiao G, Wubin H, Yue W, Jianhua H, Huachuan Z, et al. Secreted GRP78 activates EGFR-SRC-STAT3 signaling and confers the resistance to sorafeinib in HCC cells. *Oncotarget* 2017;**8**:19354–64.

81. Miharada K, Karlsson G, Rehn M, Rorby E, Siva K, Cammenga J, et al. Cripto regulates hematopoietic stem cells as a hypoxic-niche-related factor through cell surface receptor GRP78. *Cell Stem Cell* 2011;**9**:330–44.

82. de Ridder GG, Gonzalez-Gronow M, Ray R, Pizzo SV. Autoantibodies against cell surface GRP78 promote tumor growth in a murine model of melanoma. *Melanoma Res* 2011;**21**:35–43.

83. Martin-Perez R, Niwa M, Lopez-Rivas A. ER stress sensitizes cells to TRAIL through down-regulation of FLIP and Mcl-1 and PERK-dependent up-regulation of TRAIL-R2. *Apoptosis* 2012;**17**:349–63.

84. Fu Y, Lee AS. Glucose regulated proteins in cancer progression, drug resistance and immunotherapy. *Cancer Biol Ther* 2006;**5**:741–4.

85. Sun Q, Hua J, Wang Q, Xu W, Zhang J, Zhang J, et al. Expressions of GRP78 and Bax associate with differentiation, metastasis, and apoptosis in non-small cell lung cancer. *Mol Biol Rep* 2012;**39**:6753–61.

86. Daneshmand S, Quek ML, Lin E, Lee C, Cote RJ, Hawes D, et al. Glucose-regulated protein GRP78 is up-regulated in prostate cancer and correlates with recurrence and survival. *Hum Pathol* 2007;**38**:1547–52.

87. Mandelin J, Cardo-Vila M, Driessen WH, Mathew P, Navone NM, Lin SH, et al. Selection and identification of ligand peptides targeting a model of castrate-resistant osteogenic prostate cancer and their receptors. *Proc Natl Acad Sci USA* 2015;**112**:3776–81.

88. Dang CV, O'Donnell KA, Zeller KI, Nguyen T, Osthus RC, Li F. The c-Myc target gene network. *Semin Cancer Biol* 2006;**16**:253–64.

89. Sabo A, Kress TR, Pelizzola M, de Pretis S, Gorski MM, Tesi A, et al. Selective transcriptional regulation by Myc in cellular growth control and lymphomagenesis. *Nature* 2014;**511**:488–92.

90. Misra UK, Pizzo SV. Activated alpha2-macroglobulin binding to human prostate cancer cells triggers insulin-like responses. *J Biol Chem* 2015;**290**:9571–87.

91. Gopal U, Pizzo SV. Cell surface GRP78 promotes tumor cell histone acetylation through metabolic reprogramming: a mechanism which modulates the Warburg effect. *Oncotarget* 2017;**8**:107947–63.

92. Raiter A, Weiss C, Bechor Z, Ben-Dor I, Battler A, Kaplan B, et al. Activation of GRP78 on endothelial cell membranes by an ADAM15-derived peptide induces angiogenesis. *J Vasc Res* 2010;**47**:399–411.

93. Triantafilou K, Fradelizi D, Wilson K, Triantafilou M. GRP78, a coreceptor for coxsackievirus A9, interacts with major histocompatibility complex class I molecules which mediate virus internalization. *J Virol* 2002;**76**:633–43.

94. Jindadamrongwech S, Thepparit C, Smith DR. Identification of GRP 78 (BiP) as a liver cell expressed receptor element for dengue virus serotype 2. *Arch Virol* 2004;**149**:915–27.

95. Honda T, Horie M, Daito T, Ikuta K, Tomonaga K. Molecular chaperone BiP interacts with Borna disease virus glycoprotein at the cell surface. *J Virol* 2009;**83**:12622–5.

96. Chang CC, Campoli M, Ferrone S. HLA class I antigen expression in malignant cells: why does it not always correlate with CTL-mediated lysis?. *Curr Opin Immunol* 2004;**16**:644–50.

97. Marincola FM, Jaffee EM, Hicklin DJ, Ferrone S. Escape of human solid tumors from T-cell recognition: molecular mechanisms and functional significance. *Adv Immunol* 2000;**74**:181–273.

98. Misra UK, Chu CT, Rubenstein DS, Gawdi G, Pizzo SV. Receptor-recognized alpha 2-macroglobulin-methylamine elevates intracellular calcium, inositol phosphates and cyclic AMP in murine peritoneal macrophages. *Biochem J* 1993;**290**(Pt 3):885–91.

99. Misra UK, Pizzo SV. Evidence for a pro-proliferative feedback loop in prostate cancer: the role of Epac1 and COX-2-dependent pathways. *PLoS One* 2013;**8**:e63150.

100. Oida T, Weiner HL. TGF-beta induces surface LAP expression on murine CD4 T cells independent of Foxp3 induction. *PLoS One* 2010;**5**:e15523.

101. Corrigall VM, Vittecoq O, Panayi GS. Binding immunoglobulin protein-treated peripheral blood monocyte-derived dendritic cells are refractory to maturation and induce regulatory T-cell development. *Immunology* 2009;**128**:218–26.

102. Blass S, Union A, Raymackers J, Schumann F, Ungethum U, Muller-Steinbach S, et al. The stress protein BiP is overexpressed and is a major B and T cell target in rheumatoid arthritis. *Arthritis Rheum* 2001;**44**:761–71.

103. Corrigall VM, Bodman-Smith MD, Fife MS, Canas B, Myers LK, Wooley P, et al. The human endoplasmic reticulum molecular chaperone BiP is an autoantigen for rheumatoid arthritis and prevents the induction of experimental arthritis. *J Immunol* 2001;**166**:1492–8.

104. Shoda H, Fujio K, Shibuya M, Okamura T, Sumitomo S, Okamoto A, et al. Detection of autoantibodies to citrullinated BiP in rheumatoid arthritis patients and pro-inflammatory role of citrullinated BiP in collagen-induced arthritis. *Arthritis Res Ther* 2011;**13**:R191.

105. Yoshida H. ER stress and diseases. *FEBS J* 2007;**274**:630–58.

106. Corrigall VM, Bodman-Smith MD, Brunst M, Cornell H, Panayi GS. Inhibition of antigen-presenting cell function and stimulation of human peripheral blood mononuclear cells to express an antiinflammatory cytokine profile by the stress protein BiP: relevance to the treatment of inflammatory arthritis. *Arthritis Rheum* 2004;**50**:1164–71.

107. Bodman-Smith MD, Corrigall VM, Kemeny DM, Panayi GS. BiP, a putative autoantigen in rheumatoid arthritis, stimulates IL-10-producing CD8-positive T cells from normal individuals. *Rheumatology (Oxford)* 2003;**42**:637–44.

108. Brownlie RJ, Myers LK, Wooley PH, Corrigall VM, Bodman-Smith MD, Panayi GS, et al. Treatment of murine collagen-induced arthritis by the stress protein BiP via interleukin-4-producing regulatory T cells: a novel function for an ancient protein. *Arthritis Rheum* 2006;**54**:854–63.

109. Panayi GS, Corrigall VM. BiP regulates autoimmune inflammation and tissue damage. *Autoimmun Rev* 2006;**5**:140–2.

110. Vossenaar ER, Zendman AJ, van Venrooij WJ, Pruijn GJ. PAD, a growing family of citrullinating enzymes: genes, features and involvement in disease. *Bioessays* 2003;**25**:1106–18.

111. Vincent C, Nogueira L, Clavel C, Sebbag M, Serre G. Autoantibodies to citrullinated proteins: ACPA. *Autoimmunity* 2005;**38**:17–24.

112. Suzuki A, Yamada R, Yamamoto K. Citrullination by peptidylarginine deiminase in rheumatoid arthritis. *Ann N Y Acad Sci* 2007;**1108**:323–39.

113. van Venrooij WJ, van Beers JJ, Pruijn GJ. Anti-CCP antibodies: the past, the present and the future. *Nat Rev Rheumatol* 2011;**7**:391–8.

114. Casciola-Rosen LA, Anhalt G, Rosen A. Autoantigens targeted in systemic lupus erythematosus are clustered in two populations of surface structures on apoptotic keratinocytes. *J Exp Med* 1994;**179**:1317–30.

115. Weber CK, Haslbeck M, Englbrecht M, Sehnert B, Mielenz D, Graef D, et al. Antibodies to the endoplasmic reticulum-resident chaperones calnexin, BiP and Grp94 in patients with rheumatoid arthritis and systemic lupus erythematosus. *Rheumatology (Oxford)* 2010;**49**:2255–63.

116. Miao YR, Eckhardt BL, Cao Y, Pasqualini R, Argani P, Arap W, et al. Inhibition of established micrometastases by targeted drug delivery via cell surface-associated GRP78. *Clin Cancer Res* 2013;**19**:2107–16.

117. Roller C, Maddalo D. The molecular chaperone GRP78/BiP in the development of chemoresistance: mechanism and possible treatment. *Front Pharmacol* 2013;**4**:10.

118. Arap MA, Lahdenranta J, Mintz PJ, Hajitou A, Sarkis AS, Arap W, et al. Cell surface expression of the stress response chaperone GRP78 enables tumor targeting by circulating ligands. *Cancer Cell* 2004;**6**:275–84.

119. Passarella RJ, Spratt DE, van der Ende AE, Phillips JG, Wu H, Sathiyakumar V, et al. Targeted nanoparticles that deliver a sustained, specific release of Paclitaxel to irradiated tumors. *Cancer Res* 2010;**70**:4550–9.

120. Larson N, Ray A, Malugin A, Pike DB, Ghandehari H. HPMA copolymer-aminohexylgeldanamycin conjugates targeting cell surface expressed GRP78 in prostate cancer. *Pharm Res* 2010;**27**:2683–93.

121. Uckun FM, Qazi S, Ozer Z, Garner AL, Pitt J, Ma H, et al. Inducing apoptosis in chemotherapy-resistant B-lineage acute lymphoblastic leukaemia cells by targeting HSPA5, a master regulator of the anti-apoptotic unfolded protein response signalling network. *Br J Haematol* 2011;**153**:741–52.

122. Kang J, Zhao G, Lin T, Tang S, Xu G, Hu S, et al. A peptide derived from phage display library exhibits anti-tumor activity by targeting GRP78 in gastric cancer multidrug resistance cells. *Cancer Lett* 2013;**339**:247–59.

123. Delie F, Petignat P, Cohen M. GRP78-targeted nanotherapy against castrate-resistant prostate cancer cells expressing membrane GRP78. *Target Oncol* 2013;**8**:225–30.

124. Kapoor V, Dadey DY, Nguyen K, Wildman SA, Hoye K, Khudanyan A, et al. Tumor-specific binding of radiolabeled PEGylated GIRLRG peptide: a novel agent for targeting cancers. *J Nucl Med* 2016;**57**:1991–7.

125. Ferrara F, Staquicini DI, Driessen WH, D'Angelo S, Dobroff AS, Barry M, et al. Targeted molecular-genetic imaging and ligand-directed therapy in aggressive variant prostate cancer. *Proc Natl Acad Sci USA* 2016;**113**:12786–91.

126. Dobroff AS, Angelo SD', Eckhardt BL, Ferrara F, Staquicini DI, Cardo-Vila M, et al. Towards a transcriptome-based theranostic platform for unfavorable breast cancer phenotypes. *Proc Natl Acad Sci USA* 2016;**113**:12780–5.

127. Lin ML, Chen SS, Ng SH. CHM-1 suppresses formation of cell surface-associated GRP78-p85alpha complexes, inhibiting PI3K-AKT signaling and inducing apoptosis of human nasopharyngeal carcinoma cells. *Anticancer Res* 2015;**35**:5359–68.

128. de Ridder GG, Ray R, Pizzo SV. A murine monoclonal antibody directed against the carboxyl-terminal domain of GRP78 suppresses melanoma growth in mice. *Melanoma Res* 2012;**22**:225–35.

129. Lee AS. The ER chaperone and signaling regulator GRP78/BiP as a monitor of endoplasmic reticulum stress. *Methods* 2005;**35**:373–81.

130. Hensel F, Eckstein M, Rosenwald A, Brandlein S. Early development of PAT-SM6 for the treatment of melanoma. *Melanoma Res* 2013;**23**:264–75.

131. Rasche L, Duell J, Castro IC, Dubljevic V, Chatterjee M, Knop S, et al. GRP78-directed immunotherapy in relapsed or refractory multiple myeloma – results from a phase 1 trial with the monoclonal immunoglobulin M antibody PAT-SM6. *Haematologica* 2015;**100**:377–84.

132. Rasche L, Menoret E, Dubljevic V, Menu E, Vanderkerken K, Lapa C, et al. A GRP78-directed monoclonal antibody recaptures response in refractory multiple myeloma with extramedullary involvement. *Clin Cancer Res* 2016;**22**:4341–9.

133. Cohen M, Petignat P. Purified autoantibodies against glucose-regulated protein 78 (GRP78) promote apoptosis and decrease invasiveness of ovarian cancer cells. *Cancer Lett* 2011;**309**:104–9.

134. Hanahan D, Weinberg RA. Hallmarks of cancer: the next generation. *Cell* 2011;**144**:646–74.

135. Graham DK, DeRyckere D, Davies KD, Earp HS. The TAM family: phosphatidylserine sensing receptor tyrosine kinases gone awry in cancer. *Nat Rev Cancer* 2014;**14**:769–85.

Cell Surface GRP78: Anchoring and Translocation Mechanisms and Therapeutic Potential in Cancer

Yuan-Li Tsai and Amy S. Lee
USC Norris Comprehensive Cancer Center, Los Angeles, CA, United States

INTRODUCTION

Glucose-regulated protein 78 kDa (GRP78), also referred to as BiP/HSPA5, is a member of the 70 kDa heat shock protein (HSP70) family and evolutionarily conserved from yeast to humans. GRP78 contains several domains that are critical for its localization and function. For

Cell Surface GRP78, a New Paradigm in Signal Transduction Biology. DOI: https://doi.org/10.1016/B978-0-12-812351-5.00003-9
© 2018 Elsevier Inc. All rights reserved.

example, its N-terminal signal peptide targets it to the endoplasmic reticulum (ER) and the C-terminal KDEL motif marks it for ER retention. As a molecular chaperone, GRP78 contains an ATPase domain and a substrate-binding domain, which facilitate folding of nascent peptides in the ER.[1] GRP78 also acts as a key regulator of unfolded protein response (UPR). In nonstressed cells, GRP78 maintains the three transmembrane UPR sensors (PERK, IRE1 and ATF6) inactive through direct binding.[2,3] Upon ER stress, accumulated misfolded proteins titrate GRP78 away, releasing UPR sensors and leading to activation of UPR signals. GRP78 is a potent antiapoptotic protein. This could be in part due to the ability of GRP78 to form complex and sequester procaspase-7, an executioner caspase[4,5] and BIK, a proapoptotic member of the Bcl-2 family,[6] both of which are located at the outer surface of the ER.

In cancer, up-regulation of GRP78 is widely observed and associated with aggressive growth and invasiveness.[3,7,8] While GRP78 is traditionally regarded as an ER lumenal protein, studies have emerged which show that GRP78 can be detected in other cellular compartments including cell surface, cytosol, nucleus, and mitochondria.[9] Of these, cell surface GRP78 (CS-GRP78) is most studied and particularly relevant in cancer, acting both as a mediator and target of anticancer therapy.[2,3] Unlike its role in the ER in processing and folding of nascent proteins, GRP78 exhibits different functions on the cell surface, where it regulates critical oncogenic signaling pathways.[10–12] Evidence is accumulating that GRP78 translocates from the ER to the cell surface under stress typical of the tumor microenvironment, including hypoxia and glucose depletion, and cancer cells that have acquired therapeutic resistance express higher CS-GRP78 levels than the sensitive parental cells.[13–15] The discovery that GRP78 is preferentially expressed on the surface of cancer cells, but not normal organs in vivo, further opens a promising strategy for tumor specific targeting.[15–18] However, the mechanisms for stress-induced translocation of GRP78 to the cell surface are just emerging. Considering the significance of CS-GRP78 in basic cancer biology and therapeutic targeting, it is important to understand how GRP78 escapes from the ER retention machinery and as a protein without traditional transmembrane domain, stably exists on the cell surface and exerts its signaling and other biological function. In this chapter, we will discuss the current understanding of the anchoring and translocation mechanisms of

GRP78 to the cell surface, the signaling pathways it controls in concert with its binding partners and novel strategies targeting CS-GRP78 for anticancer therapy. Knowledge gained from GRP78 may also apply to other ER chaperones bearing the ER retention KDEL motif, which are also found on the cell surface in stressed cells performing critical cellular functions beyond the ER.

ANCHORING ON THE CELL SURFACE

Cell Surface Anchoring of ER-Transmembrane Chaperones

Transmembrane proteins are able to anchor on the cell surface by direct embedding into the lipid bilayer, and detection of transmembrane ER chaperones and cochaperones on the cell surface has been reported. As examples, calnexin (CNX), an ER-transmembrane chaperone important for the folding of newly synthesized glycoproteins, was detected on the surface of various cell types.[19] Using a membrane-impermeable biotinylating reagent that only labeled proteins exposed outside the cell surface, biotinylated CNX purified through binding to avidin was found to be continuously delivered to the cell surface and then endocytosed for lysosomal degradation. Murine tumor cell DnaJ-like protein 1 (MTJ-1), also referred to as DNAJC1, is an ER-localized J domain-containing transmembrane cochaperone which has been reported to partner with GRP78 and stimulates its ATPase activity in the ER.[20,21] MTJ-1 is detected in the purified plasma membrane fraction from murine macrophages and the silencing of MTJ-1 by siRNA down-regulates the expression of MTJ-1 in the cell membrane fraction.[22]

Although GRP78 was discovered as a lumenal protein in the ER, surprisingly it was observed that a subpopulation of GRP78 isolated from microsomes was resistant to alkaline carbonate extraction and partially trypsin-resistant, which are the characteristics of a transmembrane/membrane-embedded protein, while calreticulin (CRT), another ER lumenal chaperone, was sensitive to such extraction.[4,5,23] This suggests the existence of a subfraction of ER GRP78 which spans the ER membrane. However, when subjecting primary amino acid sequence of GRP78 to prediction of transmembrane helices, with the exception of the N-terminal leader peptide, the hydropathy plot only predicts three weak hydrophobic regions, all of which are below threshold of significance,[24] suggesting GRP78 does not possess transmembrane configuration. Thus, to explain the resistance to alkaline extraction and partial digestion by protease, it is possible that a subfraction of ER GRP78 "spans" the ER membrane in an unconventional fashion, or it is tightly associated with the ER membrane. Since the plasma membrane is derived from the ER membrane, the subfraction of the ER-transmembrane GRP78 could be relocated to the cell surface. In support, a minor but detectable amount of biotinylated cell surface GRP78 was found in plasma membrane fraction in stressed cells despite extensive washing to deplete peripheral proteins.[24]

Cell Surface Anchoring of Peripheral Proteins

The revelation that the majority of GRP78 on the cell surface could be washed off strongly suggests GRP78 is a peripheral protein and that anchoring of GRP78 on the cell surface is not self-dependent, but via association with other membrane proteins. To peripherally associate with the plasma membrane, proteins without transmembrane domain depend on electrostatic interactions that mediate the binding to membrane or membrane proteins. For example, Annexin A2—a member of the calcium-dependent anionic phospholipid-binding protein family—is a cytosolic protein, but has been found to peripherally associate with endosomes and cell surface.[25] The association is mediated by calcium ions which electrostatically interact with negatively charged phospholipid and residues of Annexin A2. Removal of calcium ions by EGTA results in dissociation of Annexin A2 from the cell surface.[26]

Alkaline carbonate extraction is one of the standard methods to analyze the association of protein with membrane. An extremely basic solution (pH 11.5) created by alkaline carbonate disrupts the electrostatic interactions that hold peripheral protein on the membrane, but leaves membrane-embedded proteins unaffected.[27] The subsequent ultracentrifugation precipitates lipid-embedded proteins in membrane pellet and peripheral proteins remain in supernatant. Using this method, Annexin A2 was detected in supernatant after sodium carbonate treatment, therefore it is characterized as a peripheral protein.[28]

Unlike Annexin A2 that translocates from the cytosol to the cell surface, GRP94 is an ER-originated protein which was detected on the surface of tumor cells. Commonly coregulated with GRP78, GRP94 is an abundant chaperone in the ER and was found on the surface of cancer cells using flow cytometric analysis and immunofluorescent microscopy.[29] When subjecting purified plasma membranes to sodium carbonate extraction, GRP94 was detected in aqueous supernatant, indicating cell surface GRP94 is a peripheral membrane protein. However, one of the critical challenges to analyze cell surface translocated ER chaperones is their high abundance in the ER and their low percentage on the cell surface, which might cause potential and substantial contamination in isolated cell surface fraction with the intracellular population. In 293T cells, e.g., only 4% of GRP78 in the ER pool translocated to the cell surface after ER stress induction, and 8% if GRP78 is ectopically expressed.[13] Thus stringent steps need to be taken to specifically isolate cell surface proteins to avoid possible contamination from the intracellular population, and the purity of cell surface fractions needs to be validated with markers for both the cell surface and the intracellular organelles. Established cell surface and intracellular proteins should also be examined in parallel to account for yield and recovery of each fraction during the isolation process.

A method combining cell surface protein biotinylation and alkaline carbonate extraction has been utilized to isolate and characterize CS-GRP78.[24] Surface proteins on live cells were first biotinylated, followed by sodium carbonate extraction to separate the peripheral/soluble form of GRP78 from membrane-embedded GRP78. Then fractions were subjected to avidin pull-down to isolate biotinylated CS-GRP78 from intracellular GRP78. This approach revealed that intracellularly membrane-embedded form of GRP78 was detectable only under ER stress and the majority, more than 90%, of intracellular GRP78 is soluble, presumably as an ER lumenal protein. On the cell surface, GRP78 is either undetectable, or at very low levels, in non-stressed cells. Upon ER stress, a low (<4%), but detectable membrane-embedded form of GRP78 was found, however the majority of GRP78 was detected in the soluble fraction, indicating that CS-GRP78 primarily exists as a peripheral protein.

In support of the notion that CS-GRP78 anchors stably on the cell surface via protein–protein interaction with its binding partners, cells treated with a membrane-impermeable protein crosslinker DTSSP followed by alkaline carbonate extraction showed an increase of CS-GRP78 detected in the cell membrane fraction.[24] When further analyzing the types of cell surface proteins bound to CS-GRP78 in HeLa cells, phosphoinositide phospholipase C which cleaves GPI-anchor was applied to release GPI-anchored proteins from cell surface and the conditioned medium was collected for CS-GRP78 detection. Interestingly, 30% of ectopically expressed GRP78 stably binds to GPI-anchored proteins on the cell surface. However, only 5% of endogenous CS-GRP78 associates with GPI-anchored proteins under ER stress. A possible explanation for the substantial difference is that it has been reported that ER-stress induced acute clearance of mutant, misfolded GPI-anchored proteins by targeting them to lysosomal degradation,[30] resulting in fewer GPI-anchored proteins on the cell surface that are available to partner with CS-GRP78.

The discovery that CS-GRP78 is primarily a peripheral protein supports the accessibility to antibodies recognizing C-, N-terminal, and middle region of GRP78 when externally applied.[13,31,32] The prediction of transmembrane helices based on primary amino acid sequence of GRP78 and reactivity with antibodies projected potential intracellular, extracellular, and transmembrane domains—although below the threshold of significance.[5,13,24] Referring to this predicted structure, CS-GRP78-targeting autoantibodies recognize aa 98–115 in an intracellular region, and subtilase cytotoxin (SubAB) cleaves CS-GRP78 at aa 416–417 which resides in transmembrane helix.[33,34] Thus, the recent determination that GRP78 could be entirely exposed outside the cell surface as a peripheral protein provides an explanation for the ability of the autoantibodies and bacterial toxin mentioned above to target CS-GRP78.

CS-GRP78 BINDING LIGANDS, MEMBRANE PROTEINS AND FUNCTIONS

To understand the function of CS-GRP78, it is important to identify binding partners and the pathways they control, since therapeutic agents disrupting their interactions have the potential to render those pathways inoperative. Thus far, several secreted proteins and ligands including α_2-macroglobulin (α_2M), PAR-4, Kringle 5, and isthmin are known to bind to CS-GRP78, others such as Cripto and T-cadherin are GPI-anchored cell surface proteins that bind to CS-GRP78.[9] In each case, CS-GRP78 is required for their oncogenic signaling and apoptosis. α_2M is a plasma proteinase inhibitor with broad specificity that was found to be secreted by cultured cells, and binding to proteinases induces a conformational change and activation of α_2M, designated α_2M^*.[35] GRP78 was discovered as an α_2M^* interacting protein on the plasma membrane of prostate cancer cells.[31] Blocking CS-GRP78 with anti-GRP78 antibody abolished α_2M^*-induced IP_3 production and up-regulation of free Ca^{2+} in the cytosol, $[Ca^{2+}]_i$. A further study revealed that α_2M^* can activate multiple signaling pathways including PI3K/Akt, MAPK, JNK, and NF-κB, promote c-Myc transcriptional activity, migration mediated by Pak-2/LIMK, and lipogenesis in prostate cancer.[36–41] Silencing GRP78 by RNAi or treating the cells with anti-GRP78 antibody blocks α_2M^* downstream signaling, demonstrating GRP78 as an obligatory partner of α_2M^* on the cell surface.

Another binding partner of CS-GRP78 that may help retain GRP78 on the cell surface to regulate signaling pathways is Cripto, a GPI-anchored protein playing a significant role in embryonic development, stem cell maintenance, and oncogenesis.[12,42,43] On the cell surface, GRP78 complexes with Cripto, where they cooperatively suppress TGF-β signaling by blocking Smad phosphorylation and enhancing the MAPK/PI3K pathway through activation of Src.[32,44] Coexpression of GRP78 and Cripto increases cell proliferation and colony formation of embryonal carcinoma. Down-regulating CS-GRP78 by shRNA or anti-GRP78 antibody against the N-terminal domain enhances Smad phosphorylation responding to TGF-β and Activin, and reduces activating phosphorylation of Src, Akt, ERK, and cell proliferation.[32] Thus, CS-GRP78 is a critical binding partner of Cripto for its protumorigenic activities.

T-cadherin, an atypical member of the cadherin superfamily of adhesion molecules highly expressed on vascular endothelial cells, attaches to the plasma membrane by GPI-anchor, but not transmembrane domain.[45] Expression of T-cadherin promotes cell survival by activation of Akt and GSK3β pathways. Mass spectrometric analysis revealed GRP78 as an interacting partner of T-cadherin on the cell surface.[46] Silencing of GRP78 with siRNA, or treating serum-starved cells with anti-GRP78 antibodies, blocks T-cadherin-induced Akt/GSK3β signaling, accompanied with increased procaspase-3 cleavage and loss of cell viability, validating that the interaction between T-cadherin and CS-GRP78 is crucial for T-cadherin signaling.

Other than playing a prosurvival role, CS-GRP78 has also been reported to mediate cell death. A previously identified cytoplasmic and nuclear protein, prostate apoptosis response-4 (Par-4), was discovered to be secreted out of the cell with proapoptotic activity through FADD and extrinsic apoptosis pathway.[47,48] Mass spectrometry identified GRP78 as a binding partner of Par-4, and GRP78 colocalizes with Par-4 in the ER and on the cell surface. Anti-GRP78 antibodies against N-terminal domain, but not the C-terminus, block apoptosis induced by extracellular Par-4. Expression of a cell membrane-targeting GRP78 sensitized cancer cells to Par-4 induced apoptosis, while a GRP78 mutant lacking 66 amino acids at the N-terminus did not. These studies show that CS-GRP78 is required for Par-4 proapoptotic effect in prostate cancer cells.

Kringle 5, a proteolyzed product from plasminogen, exhibits inhibitory activity to proliferating endothelial cells and cancer cells by inducing apoptosis.[49] The active Kringle 5 peptide PRKLYDY binds to CS-GRP78 as identified by mass spectrometry. Knock-down of GRP78 by siRNA results in less Kringle 5 attached to the cell surface. Incubation of human microvascular endothelial cells (HMVEC) with anti-GRP78 antibodies against N-terminus or siRNA against GRP78 potently blocked Kringle 5-induced apoptosis.[50,51]

Isthmin, another secreted protein, displays antiangiogenic activity by inducing endothelial cell apoptosis in vitro and suppresses subcutaneous B16 melanoma cell growth in mice when stably over-expressed.[52] Mass spectrometric analysis identified GRP78 as a binding protein of isthmin on the cell surface.[53] It is reported that CS-GRP78 complexed with isthmin and translocated together to mitochondria via clathrin-mediated endocytosis. Isthmin then interacted with ADP/ATP carriers and blocked ATP export from mitochondria to cytosol, resulting in apoptosis. Blocking of CS-GRP78 by anti-GRP78 antibodies against N-terminus reduced isthmin-induced apoptosis, and over-expression of GRP78 in a low basal CS-GRP78 colorectal cancer cell line LS 174T sensitized cells to isthmin-induced apoptosis, demonstrating the important role of CS-GRP78 in isthmin-triggered apoptosis.

TRANSLOCATION MECHANISMS

Intrinsic Factors

GRP78 is well established as a chaperone residing in the ER lumen. Along with other soluble ER lumenal chaperones—including CRT, protein disulfide isomerase, and GRP94—GRP78 bears a conserved C-terminal KDEL motif, which is required for its retention in the ER lumen.[54] The KDEL motif is recognized by the KDEL receptor localized in the *cis*-Golgi when ER lumenal chaperones reach Golgi complex through ER-Golgi anterograde trafficking. The binding triggers the assembly of COPI retrograde vesicle at *cis*-Golgi to deliver "escaped" chaperones back to the ER leading to ER retrieval.[55,56] It has been demonstrated that unlike the full-length protein, CRT mutant without KDEL motif is not observed in the ER, but the Golgi complex en route to being secreted in transfected cells.[57] Similarly, when the KDEL motif of GRP78 was deleted or masked by an adjacent His-tag epitope, these GRP78 mutants were detected in the culture medium, but not the full-length GRP78 with an intact KDEL motif.[13] Thus, for lumenal ER chaperones, the KDEL sequence is a critical intrinsic factor for prevention of their escape from the ER to the cell surface.

Interestingly, ectopically expressed full-length GRP78 was able to reach the cell surface in nonstressed cells and endogenous GRP78 detected on the cell surface during ER stress contained intact KDEL motif. Nonetheless, the level of KDEL receptor is not coordinately up-regulated with ER stress-induced expression of KDEL motif-containing ER chaperones.[58] Thus, the increase in GRP78 translocating from the ER to cell surface during ER stress could, in part, be due to saturation of KDEL receptor ER recycling capacity by up-regulation of ER chaperones as part of the UPR.

As a chaperone, GRP78 interacts with many client proteins via its substrate-binding domain and assists their folding through ATP hydrolysis. The ATPase domain of GRP78 was also found to mediate protein–protein interactions. Deletion of amino acids 175–201 within ATPase domain abolished GRP78 interaction with caspase-7.[5] Arginine 197, which resides in ATPase domain as well, is important for association with DnaJ cochaperones in the ER. When substituted by histidine, GRP78 R197H mutant failed to interact with DnaJ proteins.[59] An analysis of GRP78 intrinsic domains required for cell surface localization was performed with several mutants, including R197H, which is defective in cochaperone binding; G227D, which is defective in ATP binding, and T453D, which is substrate-binding defect.[60,61] While expression levels were similar among those mutants, only T453D mutant significantly reduced expression on the cell surface, indicating that substrate-binding activity is required for GRP78 translocation to the cell surface while the ATP binding appears to be dispensable.[24] Similarly to GRP78, deletion of lumenal glycoprotein binding domain of CNX blocks its cell surface delivery, suggesting the trafficking of CNX depends on the binding to glycoproteins.[19]

PROTEIN–PROTEIN INTERACTION IN GRP78 TRANSLOCATION

Consistent with the notion that CS-GRP78 translocation is facilitated through interaction with client proteins destined for the cell surface, GRP78 not only forms complexes with MTJ-1 in the ER, but was also found to coimmunoprecipitate with MTJ-1 in isolated murine macrophage plasma membrane fraction. Knock-down of MTJ-1 attenuated the expression of MTJ-1 and GRP78 in plasma membrane fraction, as well as the CS-GRP78-mediated increase of cytosolic free Ca^{2+} when stimulated with α_2M^*.[22] In parallel, the human homolog of MTJ-1, HTJ-1 was found on the surface of human endothelial cells in association with CS-GRP78, which through binding to oxidized phospholipids (OxPAPCs) promoted endothelial cell barrier enhancement.[62] Correspondingly, knock-down of HTJ-1 by siRNA significantly reduced CS-GRP78 expression and abolished OxPAPC protection against interleukin-6 and ventilator-induced lung injury.

Another reported protein carrier to help translocate GRP78 to the cell surface is Par-4. It was discovered that prostate cancer cells treated with TNF-related apoptosis-inducing ligand, TRAIL, caused ER stress and induced translocation of GRP78 from ER to the cell surface to mediate extrinsic apoptosis pathway.[48] Down-regulation of Par-4 by siRNA suppressed CS-GRP78 expression level, but had no effect on total intracellular GRP78 expression. While how intracellular Par-4 formed complex with ER GRP78 remained to be clarified,[63] GRP78 apparently interacted with Par-4 both in the ER and on the cell surface. Interestingly, in nonstressed cells, endogenous Par-4 still forms complex with GRP78, but Par-4/GRP78 complex does not translocate to the cell surface until stimulated with externally applied TRAIL or Par-4, suggesting that Par-4 alone is not sufficient to translocate GRP78 to the cell surface, but other stimulatory activities triggered by extrinsic factors also participate. Collectively, these examples illustrate that the protein binding partners of GRP78 that are responsible for its cell surface translocation could be cell context dependent, correlating with the physiological function of the cells and the type of proteins that interact with GRP78 in the ER.

TRAFFICKING ROUTES

Eukaryotic cells utilize multiple routes to secrete proteins out of the cell. The secretion of signal peptide containing proteins is initiated in the ER, where the nascent polypeptide is folded and modified, followed by exit via ER exit site (ERES) in a COPII-dependent manner.[64] Most of the vesicle derived from ER is delivered to Golgi complex before reaching final destination for further modification and segregation, and this is considered the conventional secretory pathway. However, some proteins were discovered to traffic independently of this conventional pathway, and thus are classified into unconventional secretory pathway. Cytosolic proteins such as HSP70 and Annexin A2 translocate to the cell surface from the cytosol via unconventional mechanisms.[65,66] Some proteins contain signal peptide for ER entry, but exit the ER independently of COPII-coated vesicle. ER degradation enhancing mannosidase-like 1 (EDEM1) is a key enzyme that removes the mannose groups from the terminally misfolded glycoproteins and targets them for ER-associated degradation.[67] In nonstressed cells, to minimize the interruption of ongoing folding processes by EDEM1, EDEM1 is transported to late endosome/lysosome for degradation.[68] However, its vesicle buds occur in rough ER, but not ERES, and those vesicles containing EDEM1 lack COPII coat,[69] implying EDEM1 transportation out of ER is via a COPII-independent manner. A third group of proteins traffics to the cell surface bypassing the Golgi complex. The most well-studied example is cystic fibrosis transmembrane conductance regulator (CFTR). Although CFTR exits ER in a COPII-dependent manner, it is insensitive to loss of Syntaxin 5 and Rab 1A/Rab 2 GTPases, which are required components for conventional ER-to-Golgi trafficking.[70] Rather, the expression of CFTR on the plasma membrane is suppressed by over-expression of dominant negative Syntaxin 13, a *t*-SNARE localized in *trans*-Golgi network and recycling endosomes, demonstrating that the transport of CFTR from ER to the cell surface is mediated by endosomal systems.[70]

To determine whether the trafficking route of a molecule is Golgi-dependent or Golgi-bypass, brefeldin A (BFA) treatment is commonly used.[64] BFA is a fungal metabolite that inhibits GTP-exchange factors of Arf1, a critical small GTPase that regulates COPI recruitment to Golgi membrane and vesicle formation for Golgi-to-ER retrograde.[71] Blocking COPI coats from binding to Golgi membranes results in tubulation of Golgi and eventually the absorption of the Golgi into the ER.[72] Therefore, proteins that can be secreted out of the cell upon treatment with BFA are likely to use an unconventional secretory pathway independent of ER-to-Golgi transport.

Further studies on the translocation of GRP78 to the cell surface revealed that different cell lines can use different routes for ER-stress induced CS-GRP78 expression.[24] The dependence on Golgi integrity was tested in HeLa (the cervical cancer cell line) and HCT116 (the colon cancer cell line) treated with thapsigargin or tunicamycin, which induced ER stress by blocking ER calcium pump or inhibiting N-linked protein glycosylation, respectively. As expected, BFA efficiently blocked cell surface expression of EphB4, a transmembrane receptor tyrosine kinase that is known to localize to the plasma membrane via the ER-Golgi trafficking route, and had no effect on cell surface expression of Annexin A2, whose translocation to the cell surface is independent of Golgi complex.[65] ER stress-induced CS-GRP78 expression was suppressed in HeLa cells by BFA, but, unexpectedly, not in HCT116 cells.[24] With the identification of GRP78 in exosomes, this may represent a novel pathway for transporting intracellular GRP78 outside the ER, onto the cell surface followed by secretion.[73]

THERAPEUTIC POTENTIAL IN CANCER

The preferential expression of GRP78 on the surface of cancer cells, but not normal organs in vivo, opens up the possibility of utilizing CS-GRP78 for cancer-specific therapy with minimal damaging effect on normal cells.[3] Toxicity to the tumor cells can be achieved by either treating the cancer cells with antagonistic antibodies against CS-GRP78, thereby suppressing the CS-GRP78 oncogenic functions, or utilize CS-GRP78 as a cell surface portal for the delivery of cytotoxic agents. Another approach is to apply proapoptotic ligands of CS-GRP78 which can trigger cell death. Since both tumor and tumor-associated endothelial cells, and possibly other stroma cells recruited to the tumor sites expressing high levels of CS-GRP78 under stress conditions are commonly observed in the tumor microenvironment, the cytotoxic effect dually affects the tumor and tumor-supporting cells that undergo ER-stress induced CS-GRP78 up-regulation.

ANTIBODIES

PAT-SM6, a human monoclonal IgM antibody, was isolated from a gastric cancer patient with high affinity to GRP78 and relatively weaker affinity to low-density lipoprotein (LDL).[74,75] The multiple and simultaneous bindings of PAT-SM6 to CS-GRP78 and LDL induce receptor-mediated endocytosis leading to accumulation of intracellular neutral lipids and apoptosis.[75–77] A moderate binding of PAT-SM6 to complement C1q was also reported, whose recruitment and deposition to the plasma membrane of cancer cells by PAT-SM6 triggered complement dependent cytotoxicity and partially accounted for the cell death induced by PAT-SM6 in multiple myeloma (MM) cell lines and CD138-purified primary MM cells.[78] It was also reported that combinational treatment of anti-MM agents, dexamethasone, lenalidomide, or bortezomib with PAT-SM6 had synergistic growth inhibition in sensitive and resistant cell lines in vitro. A patient with triple resistant MM received a combination of PAT-SM6, bortezomib, and lenalidomide and showed partial regression in MM lesions.[79] In the Phase I trial, single-agent PAT-SM6 in MM patients was well-tolerated and resulted in stable disease in 33% of patients.[80] Cell proliferative suppression and apoptosis induced by PAT-SM6 was observed in melanoma cell lines as well, and treatment of PAT-SM6 in human C8161 melanoma metastasis model in vivo showed a decrease in the number of lung metastases.[81]

MAb159 is a mouse monoclonal anti-GRP78 IgG antibody that displayed high specificity and affinity to GRP78 and triggered endocytosis of CS-GRP78.[15] MAb159 was found to suppress a wide range of xenograft growth of solid tumors, alone or in combination with standard therapy. MAb159 treatment resulted in inhibition of cell proliferation and the PI3K/Akt pathway without compensatory activation of the MAPK pathway in the tumors. MAb159 is able to block breast cancer 4T1 xenograft model of tumor metastasis to lung and tumor progression in spontaneous *PTEN*–loss-driven prostate and leukemia tumor models. Humanized MAb159 has been developed which retained affinity to GRP78 and cytotoxic activities to cancer.

In another study, C107 mouse monoclonal IgG antibody with high affinity to GRP78 was generated.[82] C107 was able to induce apoptosis in B16F1 murine melanoma cells in vitro. In the xenograft mouse model of B16F1 melanoma, C107 pretreatment and postadministration significantly attenuated tumor growth and extended overall survival.[82]

PEPTIDE CONJUGATES

The efficacy of a phage display-selected peptide WDLAWMFRLPVG that specifically targets GRP78 was investigated when conjugated to anticancer agents aminohexylgeldanamycin (AHGDM), docetaxel (DOC), or cisplatin.[16,83] Peptide-DOC displayed the highest antitumor activity and all the peptide conjugates exhibited higher potency in growth inhibition than the unconjugated in vitro using the human prostate cancer cell line DU145. When combined with tumor hyperthermia, AHGDM and DOC conjugates demonstrated significant synergistic cytotoxicity. Single-agent treatment with peptide-DOC conjugate showed higher tumor suppressive efficacy than DOC alone in vivo, but peptide-DOC combined with hyperthermia resulted in the strongest tumor regression in DU145 tumor bearing mice.[83] Conjugating the peptide to D-$(KLAKLAK)_2$, an apoptosis-inducing oligopeptide, resulted in reduced cell viability in vitro and tumor size in the xenograft mouse model.[16]

Peptide WIFPWIQL, identified together with WDLAWMFRLPVG, showed high specificity to GRP78 and tumor cells in vitro, reduced cell viability dependent of receptor-mediated internalization when conjugated to D-$(KLAKLAK)_2$.[16] The peptide conjugate suppressed prostate and breast tumor growth in xenografts. The anticancer effect of WIFPWIQL-conjugated liposomes was also investigated.[84] GRP78 in the isolated membrane fraction was found to be up-regulated in VEGF-stimulated HUVEC cells. The WIFPWIQL-liposome conjugate was highly taken up in vitro and in vivo by tumor-associated endothelial cells. The doxorubicin-incorporated WIFPWIQL-liposome inhibited angiogenesis in the tumor area and suppressed colon tumor growth in xenograft mouse models. Growth inhibitory activity was reported using *N*-(2-hydroxypropyl)methacrylamide copolymer conjugated to geldanamycin and WIFPWIQL in human prostate cancer cells in vitro and WIFPWIQL-micelles conjugate encapsulated in doxorubicin in human gastric cancer in vitro and xenograft mouse models.[85,86] A further study constructed a fusion protein consisting of WIFPWIQL and an active fragment of mung bean trypsin inhibitor which exhibited antiproliferative and proapoptotic effects specific in cancer cells.[87] It showed that the fusion protein induced G1 phase arrest and activated intrinsic, extrinsic, and ER stress associated apoptosis pathways in vitro and in xenograft mouse models of human colorectal carcinoma.

Cyclic 13-mer Pep42, CTVALPGGYVRVC, discovered from phage display, was found to bind to highly metastatic melanoma cell line Me6652/4, but not low metastatic Me6652/56, by recognizing CS-GRP78 identified by mass spectrometry.[88] Pep42 selectively interacts with CS-GRP78 and triggers clathrin-mediated endocytosis of pep42/GRP78 complex and delivery to lysosome.[89] Pep42-Taxol conjugate exhibited enhanced anticancer activity to melanoma. Induction of apoptosis was also observed when pep42 was conjugated to D-$(KLAKLAK)_2$ or hematoporphyrin, a photosensitizer, in high CS-GRP78 cancer cells Me6652/4 and osteosarcoma cell line SJSA-1, but not CS-GRP78 negative normal human dermal fibroblasts and lung fibroblasts. In vivo xenograft mouse models demonstrated that following intradermal injection, Pep42-conjugate was enriched in tumor tissue, but not in normal organs.[89]

Peptide GIRLRG was identified from phage display that recognized irradiated tumor lesions. Mass spectrometric analysis identified GRP78 as the target of GIRLRG which was highly expressed after irradiation therapy. In heterotopic tumor mouse models injected with murine glioma or human breast cancer cells, GIRLRG-conjugated nanoparticles encapsulated in paclitaxel results in elevated paclitaxel concentration and apoptosis in irradiated tumor lesions, significantly impeding tumor growth in vivo.[90]

PROAPOPTOTIC LIGANDS

Kringle 5 is reported to induce endothelial and cancer cell apoptosis through CS-GRP78.[50] In liver metastasis cancer models using human colorectal cancer cell line, injection with recombinant Kringle 5-induced apoptosis in tumor-associated endothelial cells and suppressed liver metastasis. Combined treatment of Kringle 5 with 5-fluorouracil (5-FU), compared with single agent blocked more liver metastasis and improved overall survival.[51] Analysis showed Kringle 5 caused cleavage of executioner procaspase-3 and -7, and initiator procaspase-9, but not procaspase-8, indicating Kringle 5-induced cell death is mediated by the intrinsic apoptotic pathway.[50,51,91]

CS-GRP78 is a high affinity receptor for isthmin, a secreted protein that induces endothelial cell apoptosis which also exhibits proapoptotic activity to cancer cells.[53] In syngeneic tumor models of melanoma and breast carcinoma, treating mice with recombinant isthmin caused increase of apoptosis and suppression of proliferation both in cancer cells and endothelial cells, accompanied by reduction of tumor vascularization and surrounding blood vessels.[53] Investigation indicated that isthmin binding to CS-GRP78 triggered clathrin-dependent endocytosis, which was blocked by Chlorpromazine. Sequentially isthmin/GRP78 complex translocated to mitochondria to inhibit ATP exchange activity of ADP/ATP carrier 2 and 3 leading to cell death.[53]

CONCLUSIONS AND PERSPECTIVES

Despite the basic science advances and the promise of CS-GRP78 in anticancer therapy, how GRP78 and ER chaperones in general traffic to the cell surface, how they are anchored on the cell surface and the signaling pathways they control, remain largely unexplored. It will be important to understand how common the Golgi-bypass route of GRP78 translocation is in cancer and its detailed mechanism, as well as how the differences of trafficking routes may affect strategies to block CS-GRP78 expression in different forms of cancer. To fully understand CS-GRP78 function in oncogenesis, it is necessary to identify its binding partners and pathways that mediate the tumorigenic process and development of therapeutic resistance. The identification of additional agents capable of inducing endocytosis and destruction of CS-GRP78, or interfering with the translocation of CS-GRP78 and other prooncogenic ER chaperones to the cell surface, warrants further research and effort.

ACKNOWLEDGMENT

This work is supported by the US National Institutes of Health (NIH) grants R01 CA027607 and R21 CA179273 and the Freeman endowed chair to A.S.L.

REFERENCES

1. Ni M, Lee AS. ER chaperones in mammalian development and human diseases. *FEBS Lett* 2007;**581**:3641–51.
2. Luo B, Lee AS. The critical roles of endoplasmic reticulum chaperones and unfolded protein response in tumorigenesis and anticancer therapies. *Oncogene* 2013;**32**:805–18.
3. Lee AS. Glucose-regulated proteins in cancer: molecular mechanisms and therapeutic potential. *Nat Rev Cancer* 2014;**14**:263–76.
4. Rao RV, Peel A, Logvinova A, del Rio G, Hermel E, Yokota T, et al. Coupling endoplasmic reticulum stress to the cell death program: role of the ER chaperone GRP78. *FEBS Lett* 2002;**514**:122–8.
5. Reddy RK, Mao C, Baumeister P, Austin RC, Kaufman RJ, Lee AS. Endoplasmic reticulum chaperone protein GRP78 protects cells from apoptosis induced by topoisomerase inhibitors: role of ATP binding site in suppression of caspase-7 activation. *J Biol Chem* 2003;**278**:20915–24.
6. Zhou H, Zhang Y, Fu Y, Chan L, Lee AS. Novel mechanism of anti-apoptotic function of 78-kDa glucose-regulated protein (GRP78): endocrine resistance factor in breast cancer, through release of B-cell lymphoma 2 (BCL-2) from BCL-2-interacting killer (BIK). *J Biol Chem* 2011;**286**:25687–96.
7. Li J, Lee AS. Stress induction of GRP78/BiP and its role in cancer. *Curr Mol Med* 2006;**6**:45–54.
8. Lee AS. GRP78 induction in cancer: therapeutic and prognostic implications. *Cancer Res* 2007;**67**:3496–9.
9. Ni M, Zhang Y, Lee AS. Beyond the endoplasmic reticulum: atypical GRP78 in cell viability, signalling and therapeutic targeting. *Biochem J* 2011;**434**:181–8.
10. Gonzalez-Gronow M, Selim MA, Papalas J, Pizzo SV. GRP78: a multifunctional receptor on the cell surface. *Antioxid Redox Signal* 2009;**11**:2299–306.
11. Sato M, Yao VJ, Arap W, Pasqualini R. GRP78 signaling hub a receptor for targeted tumor therapy. *Adv Genet* 2010;**69**:97–114.
12. Gray PC, Vale W. Cripto/GRP78 modulation of the TGF-beta pathway in development and oncogenesis. *FEBS Lett* 2012;**586**:1836–45.
13. Zhang Y, Liu R, Ni M, Gill P, Lee AS. Cell surface relocalization of the endoplasmic reticulum chaperone and unfolded protein response regulator GRP78/BiP. *J Biol Chem* 2010;**285**:15065–75.
14. Zhang Y, Tseng CC, Tsai YL, Fu X, Schiff R, Lee A. Cancer cells resistant to therapy promote cell surface relocalization of GRP78 which complexes with PI3K and enhances PI (3,4,5)P3 production. *PLoS One* 2013;**8**:e80071.
15. Liu R, Li X, Gao W, Zhou Y, Wey S, Mitra SK, et al. Monoclonal antibody against cell surface GRP78 as a novel agent in suppressing PI3K/AKT signaling, tumor growth and metastasis. *Clin Cancer Res* 2013;**19**:6802–11.
16. Arap MA, Lahdenranta J, Mintz PJ, Hajitou A, Sarkis AS, Arap W, et al. Cell surface expression of the stress response chaperone GRP78 enables tumor targeting by circulating ligands. *Cancer Cell* 2004;**6**:275–84.
17. Jakobsen CG, Rasmussen N, Laenkholm AV, Ditzel HJ. Phage display derived human monoclonal antibodies isolated by binding to the surface of live primary breast cancer cells recognize GRP78. *Cancer Res* 2007;**67**:9507–17.
18. Yoneda Y, Steiniger SC, Capkova K, Mee JM, Liu Y, Kaufmann GF, et al. A cell-penetrating peptidic GRP78 ligand for tumor cell-specific prodrug therapy. *Bioorg Med Chem Lett* 2008;**18**:1632–6.

19. Okazaki Y, Ohno H, Takase K, Ochiai T, Saito T. Cell surface expression of calnexin, a molecular chaperone in the endoplasmic reticulum. *J Biol Chem* 2000;**275**:35751–8.

20. Chevalier M, Rhee H, Elguindi EC, Blond SY. Interaction of murine BiP/GRP78 with the DnaJ homologue MTJ1. *J Biol Chem* 2000;**275**:19620–7.

21. Dudek J, Volkmer J, Bies C, Guth S, Muller A, Lerner M, et al. A novel type of co-chaperone mediates transmembrane recruitment of DnaK-like chaperones to ribosomes. *EMBO J* 2002;**21**:2958–67.

22. Misra UK, Gonzalez-Gronow M, Gawdi G, Pizzo SV. The role of MTJ-1 in cell surface translocation of GRP78, a receptor for alpha 2-macroglobulin-dependent signaling. *J Immunol* 2005;**174**:2092–7.

23. Zhang Y, Nijbroek G, Sullivan ML, McCracken AA, Watkins SC, Michaelis S, et al. Hsp70 molecular chaperone facilitates endoplasmic reticulum-associated protein degradation of cystic fibrosis transmembrane conductance regulator in yeast. *Mol Biol Cell* 2001;**12**:1303–14.

24. Tsai YL, Zhang Y, Tseng CC, Stanciauskas R, Pinaud F, Lee AS. Characterization and mechanism of stress-induced translocation of 78-kilodalton glucose-regulated protein (GRP78) to the cell surface. *J Biol Chem* 2015;**290**:8049–64.

25. Gerke V, Creutz CE, Moss SE. Annexins: linking Ca2 + signalling to membrane dynamics. *Nat Rev Mol Cell Biol* 2005;**6**:449–61.

26. Valapala M, Maji S, Borejdo J, Vishwanatha JK. Cell surface translocation of annexin A2 facilitates glutamate-induced extracellular proteolysis. *J Biol Chem* 2014;**289**:15915–26.

27. Okamoto T, Schwab RB, Scherer PE, Lisanti MP. Analysis of the association of proteins with membranes. *Curr Protoc Cell Biol* 2001. **Chapter 5**, 5:5.4:5.4.1–5.4.17.

28. Liu L, Tao JQ, Zimmerman UJ. Annexin II binds to the membrane of A549 cells in a calcium-dependent and calcium-independent manner. *Cell Signal* 1997;**9**:299–304.

29. Altmeyer A, Maki RG, Feldweg AM, Heike M, Protopopov VP, Masur SK, et al. Tumor-specific cell surface expression of the-KDEL containing, endoplasmic reticular heat shock protein gp96. *Int J Cancer* 1996;**69**:340–9.

30. Satpute-Krishnan P, Ajinkya M, Bhat S, Itakura E, Hegde RS, Lippincott-Schwartz J. ER stress-induced clearance of misfolded GPI-anchored proteins via the secretory pathway. *Cell* 2014;**158**:522–33.

31. Misra UK, Gonzalez-Gronow M, Gawdi G, Hart JP, Johnson CE, Pizzo SV. The role of Grp 78 in alpha 2-macroglobulin-induced signal transduction. Evidence from RNA interference that the low density lipoprotein receptor-related protein is associated with, but not necessary for, GRP 78-mediated signal transduction. *J Biol Chem* 2002;**277**:42082–7.

32. Kelber JA, Panopoulos AD, Shani G, Booker EC, Belmonte JC, Vale WW, et al. Blockade of Cripto binding to cell surface GRP78 inhibits oncogenic Cripto signaling via MAPK/PI3K and Smad2/3 pathways. *Oncogene* 2009;**28**:2324–36.

33. Al-Hashimi AA, Caldwell J, Gonzalez-Gronow M, Pizzo SV, Aboumrad D, Pozza L, et al. Binding of anti-GRP78 autoantibodies to cell surface GRP78 increases tissue factor procoagulant activity via the release of calcium from endoplasmic reticulum stores. *J Biol Chem* 2010;**285**:28912–23.

34. Ray R, de Ridder GG, Eu JP, Paton AW, Paton JC, Pizzo SV. The *Escherichia coli* subtilase cytotoxin A subunit specifically cleaves cell-surface GRP78 protein and abolishes COOH-terminal-dependent signaling. *J Biol Chem* 2012;**287**:32755–69.

35. Petersen CM, Christiansen BS, Jensen PH, Moestrup SK, Gliemann J, Sottrup-Jensen L, et al. Human hepatocytes exhibit receptors for alpha 2-macroglobulin and pregnancy zone protein-proteinase complexes. *Eur J Clin Invest* 1988;**18**:184–90.

36. Misra UK, Gonzalez-Gronow M, Gawdi G, Wang F, Pizzo SV. A novel receptor function for the heat shock protein Grp78: silencing of Grp78 gene expression attenuates alpha2M*-induced signalling. *Cell Signal* 2004;**16**:929–38.
37. Misra UK, Deedwania R, Pizzo SV. Binding of activated alpha2-macroglobulin to its cell surface receptor GRP78 in 1-LN prostate cancer cells regulates PAK-2-dependent activation of LIMK. *J Biol Chem* 2005;**280**:26278–86.
38. Misra UK, Deedwania R, Pizzo SV. Activation and cross-talk between Akt, NF-{kappa}B, and unfolded protein response signaling in 1-LN prostate cancer cells consequent to ligation of cell surface-associated GRP78. *J Biol Chem* 2006;**281**:13694–707.
39. Misra UK, Pizzo SV. Evidence for a pro-proliferative feedback loop in prostate cancer: the role of Epac1 and COX-2-dependent pathways. *PLoS One* 2013;**8**:e63150.
40. Misra UK, Pizzo SV. Activated alpha2-macroglobulin binding to human prostate cancer cells triggers insulin-like responses. *J Biol Chem* 2015;**290**:9571–87.
41. Gopal U, Gonzalez-Gronow M, Pizzo SV. Activated alpha2-macroglobulin regulates transcriptional activation of c-Myc target genes through cell surface GRP78. *J Biol Chem* 2016;**291**:10904–15.
42. Strizzi L, Bianco C, Normanno N, Salomon D. Cripto-1: a multifunctional modulator during embryogenesis and oncogenesis. *Oncogene* 2005;**24**:5731–41.
43. Spike BT, Kelber JA, Booker E, Kalathur M, Rodewald R, Lipianskaya J, et al. CRIPTO/GRP78 signaling maintains fetal and adult mammary stem cells ex vivo. *Stem Cell Rep* 2014;**2**:427–39.
44. Shani G, Fischer WH, Justice NJ, Kelber JA, Vale W, Gray PC. GRP78 and Cripto form a complex at the cell surface and collaborate to inhibit transforming growth factor beta signaling and enhance cell growth. *Mol Cell Biol* 2008;**28**:666–77.
45. Angst BD, Marcozzi C, Magee AI. The cadherin superfamily: diversity in form and function. *J Cell Sci* 2001;**114**:629–41.
46. Philippova M, Ivanov D, Joshi MB, Kyriakakis E, Rupp K, Afonyushkin T, et al. Identification of proteins associating with glycosylphosphatidylinositol- anchored T-cadherin on the surface of vascular endothelial cells: role for Grp78/BiP in T-cadherin-dependent cell survival. *Mol Cell Biol* 2008;**28**:4004–17.
47. El-Guendy N, Zhao Y, Gurumurthy S, Burikhanov R, Rangnekar VM. Identification of a unique core domain of par-4 sufficient for selective apoptosis induction in cancer cells. *Mol Cell Biol* 2003;**23**:5516–25.
48. Burikhanov R, Zhao Y, Goswami A, Qiu S, Schwarze SR, Rangnekar VM. The tumor suppressor Par-4 activates an extrinsic pathway for apoptosis. *Cell* 2009;**138**:377–88.
49. Cao Y, Ji RW, Davidson D, Schaller J, Marti D, Sohndel S, et al. Kringle domains of human angiostatin. Characterization of the anti-proliferative activity on endothelial cells. *J Biol Chem* 1996;**271**:29461–7.
50. Davidson DJ, Haskell C, Majest S, Kherzai A, Egan DA, Walter KA, et al. Kringle 5 of human plasminogen induces apoptosis of endothelial and tumor cells through surface-expressed glucose-regulated protein 78. *Cancer Res* 2005;**65**:4663–72.
51. Ahn JH, Yu HK, Lee HJ, Hong SW, Kim SJ, Kim JS. Suppression of colorectal cancer liver metastasis by apolipoprotein(a) kringle V in a nude mouse model through the induction of apoptosis in tumor-associated endothelial cells. *PLoS One* 2014;**9**:e93794.
52. Xiang W, Ke Z, Zhang Y, Cheng GH, Irwan ID, Sulochana KN, et al. Isthmin is a novel secreted angiogenesis inhibitor that inhibits tumour growth in mice. *J Cell Mol Med* 2011;**15**:359–74.

53. Chen M, Zhang Y, Yu VC, Chong YS, Yoshioka T, Ge R. Isthmin targets cell-surface GRP78 and triggers apoptosis via induction of mitochondrial dysfunction. *Cell Death Differ* 2014;**21**:797–810.

54. Munro S, Pelham HR. A C-terminal signal prevents secretion of luminal ER proteins. *Cell* 1987;**48**:899–907.

55. Semenza JC, Hardwick KG, Dean N, Pelham HR. ERD2, a yeast gene required for the receptor-mediated retrieval of luminal ER proteins from the secretory pathway. *Cell* 1990;**61**:1349–57.

56. D'Souza-Schorey C, Chavrier P. ARF proteins: roles in membrane traffic and beyond. *Nat Rev Mol Cell Biol* 2006;**7**:347–58.

57. Raykhel I, Alanen H, Salo K, Jurvansuu J, Nguyen VD, Latva-Ranta M, et al. A molecular specificity code for the three mammalian KDEL receptors. *J Cell Biol* 2007;**179**:1193–204.

58. Llewellyn DH, Roderick HL, Rose S. KDEL receptor expression is not coordinatedly up-regulated with ER stress-induced reticuloplasmin expression in HeLa cells. *Biochem Biophys Res Commun* 1997;**240**:36–40.

59. Awad W, Estrada I, Shen Y, Hendershot LM. BiP mutants that are unable to interact with endoplasmic reticulum DnaJ proteins provide insights into interdomain interactions in BiP. *Proc Natl Acad Sci USA* 2008;**105**:1164–9.

60. Wei J, Gaut JR, Hendershot LM. *In vitro* dissociation of BiP-peptide complexes requires a conformational change in BiP after ATP binding but does not require ATP hydrolysis. *J Biol Chem* 1995;**270**:26677–82.

61. Guo F, Snapp EL. ERdj3 regulates BiP occupancy in living cells. *J Cell Sci* 2013;**126**:1429–39.

62. Birukova AA, Singleton PA, Gawlak G, Tian X, Mirzapoiazova T, Mambetsariev B, et al. GRP78 is a novel receptor initiating a vascular barrier protective response to oxidized phospholipids. *Mol Biol Cell* 2014;**25**:2006–16.

63. Lee AS. The Par-4-GRP78 TRAIL, more twists and turns. *Cancer Biol Ther* 2009;**8**:2103–5.

64. Grieve AG, Rabouille C. Golgi bypass: skirting around the heart of classical secretion. *Cold Spring Harb Perspect Biol* 2011;**3**:a005298.

65. Deora AB, Kreitzer G, Jacovina AT, Hajjar KA. An annexin 2 phosphorylation switch mediates p11-dependent translocation of annexin 2 to the cell surface. *J Biol Chem* 2004;**279**:43411–18.

66. Liu C, Qu L, Lian S, Tian Z, Zhao C, Meng L, et al. Unconventional secretion of synuclein-gamma promotes tumor cell invasion. *FEBS J* 2014;**281**:5159–71.

67. Olivari S, Cali T, Salo KE, Paganetti P, Ruddock LW, Molinari M. EDEM1 regulates ER-associated degradation by accelerating de-mannosylation of folding-defective polypeptides and by inhibiting their covalent aggregation. *Biochem Biophys Res Commun* 2006;**349**:1278–84.

68. Cali T, Galli C, Olivari S, Molinari M. Segregation and rapid turnover of EDEM1 by an autophagy-like mechanism modulates standard ERAD and folding activities. *Biochem Biophys Res Commun* 2008;**371**:405–10.

69. Zuber C, Cormier JH, Guhl B, Santimaria R, Hebert DN, Roth J. EDEM1 reveals a quality control vesicular transport pathway out of the endoplasmic reticulum not involving the COPII exit sites. *Proc Natl Acad Sci USA* 2007;**104**:4407–12.

70. Yoo JS, Moyer BD, Bannykh S, Yoo HM, Riordan JR, Balch WE. Non-conventional trafficking of the cystic fibrosis transmembrane conductance regulator through the early secretory pathway. *J Biol Chem* 2002;**277**:11401–9.

71. Popoff V, Adolf F, Brugger B, Wieland F. COPI budding within the Golgi stack. *Cold Spring Harb Perspect Biol* 2011;**3**:a005231.

72. Lippincott-Schwartz J, Yuan LC, Bonifacino JS, Klausner RD. Rapid redistribution of Golgi proteins into the ER in cells treated with brefeldin A: evidence for membrane cycling from Golgi to ER. *Cell* 1989;**56**:801–13.

73. Xiao D, Ohlendorf J, Chen Y, Taylor DD, Rai SN, Waigel S, et al. Identifying mRNA, microRNA and protein profiles of melanoma exosomes. *PLoS One* 2012;**7**:e46874.

74. Rauschert N, Brandlein S, Holzinger E, Hensel F, Muller-Hermelink HK, Vollmers HP. A new tumor-specific variant of GRP78 as target for antibody-based therapy. *Lab Invest* 2008;**88**:375–86.

75. Rosenes Z, Mulhern TD, Hatters DM, Ilag LL, Power BE, Hosking C, et al. The anti-cancer IgM monoclonal antibody PAT-SM6 binds with high avidity to the unfolded protein response regulator GRP78. *PLoS One* 2012;**7**:e44927.

76. Pohle T, Brandlein S, Ruoff N, Muller-Hermelink HK, Vollmers HP. Lipoptosis: tumor-specific cell death by antibody-induced intracellular lipid accumulation. *Cancer Res* 2004;**64**:3900–6.

77. Brandlein S, Rauschert N, Rasche L, Dreykluft A, Hensel F, Conzelmann E, et al. The human IgM antibody SAM-6 induces tumor-specific apoptosis with oxidized low-density lipoprotein. *Mol Cancer Ther* 2007;**6**:326–33.

78. Rasche L, Duell J, Morgner C, Chatterjee M, Hensel F, Rosenwald A, et al. The natural human IgM antibody PAT-SM6 induces apoptosis in primary human multiple myeloma cells by targeting heat shock protein GRP78. *PLoS One* 2013;**8**:e63414.

79. Rasche L, Menoret E, Dubljevic V, Menu E, Vanderkerken K, Lapa C, et al. A GRP78-directed monoclonal antibody recaptures response in refractory multiple myeloma with extramedullary involvement. *Clin Cancer Res* 2016;**22**:4341–9.

80. Rasche L, Duell J, Castro IC, Dubljevic V, Chatterjee M, Knop S, et al. GRP78-directed immunotherapy in relapsed or refractory multiple myeloma – results from a phase 1 trial with the monoclonal immunoglobulin M antibody PAT-SM6. *Haematologica* 2015;**100**:377–84.

81. Hensel F, Eckstein M, Rosenwald A, Brandlein S. Early development of PAT-SM6 for the treatment of melanoma. *Melanoma Res* 2013;**23**:264–75.

82. de Ridder GG, Ray R, Pizzo SV. A murine monoclonal antibody directed against the carboxyl-terminal domain of GRP78 suppresses melanoma growth in mice. *Melanoma Res* 2012;**22**:225–35.

83. Larson N, Gormley A, Frazier N, Ghandehari H. Synergistic enhancement of cancer therapy using a combination of heat shock protein targeted HPMA copolymer-drug conjugates and gold nanorod induced hyperthermia. *J Control Release* 2013;**170**:41–50.

84. Katanasaka Y, Ishii T, Asai T, Naitou H, Maeda N, Koizumi F, et al. Cancer antineovascular therapy with liposome drug delivery systems targeted to BiP/GRP78. *Int J Cancer* 2010;**127**:2685–98.

85. Larson N, Ray A, Malugin A, Pike DB, Ghandehari H. HPMA copolymer-aminohexylgeldanamycin conjugates targeting cell surface expressed GRP78 in prostate cancer. *Pharm Res* 2010;**27**:2683–93.

86. Cheng CC, Lu N, Peng CL, Chang CC, Mai FD, Chen LY, et al. Targeting to overexpressed glucose-regulated protein 78 in gastric cancer discovered by 2D DIGE improves the diagnostic and therapeutic efficacy of micelles-mediated system. *Proteomics* 2012;**12**:2584–97.

87. Li Z, Zhao C, Zhao Y, Shan S, Shi T, Li J. Reconstructed mung bean trypsin inhibitor targeting cell surface GRP78 induces apoptosis and inhibits tumor growth in colorectal cancer. *Int J Biochem Cell Biol* 2014;**47**:68–75.

88. Kim Y, Lillo AM, Steiniger SC, Liu Y, Ballatore C, Anichini A, et al. Targeting heat shock proteins on cancer cells: selection, characterization, and cell-penetrating properties of a peptidic GRP78 ligand. *Biochemistry* 2006;**45**:9434–44.

89. Liu Y, Steiniger SC, Kim Y, Kaufmann GF, Felding-Habermann B, Janda KD. Mechanistic studies of a peptidic GRP78 ligand for cancer cell-specific drug delivery. *Mol Pharm* 2007;**4**:435–47.

90. Passarella RJ, Spratt DE, van der Ende AE, Phillips JG, Wu H, Sathiyakumar V, et al. Targeted nanoparticles that deliver a sustained, specific release of Paclitaxel to irradiated tumors. *Cancer Res* 2010;**70**:4550–9.

91. McFarland BC, Stewart Jr. J, Hamza A, Nordal R, Davidson DJ, Henkin J, et al. Plasminogen Kringle 5 induces apoptosis of brain microvessel endothelial cells: sensitization by radiation and requirement for GRP78 and LRP1. *Cancer Res* 2009;**69**:5537–45.

CHAPTER 4

Cell Surface GRP78: A Novel Regulator of Tissue Factor Procoagulant Activity

Ali A. Al-Hashimi[1], Janusz Rak[2] and Richard C. Austin[1]

[1]St. Joseph's Healthcare Hamilton and McMaster University, Hamilton, ON, Canada

[2]McGill University, Montreal, QC, Canada

Cell Surface GRP78, a New Paradigm in Signal Transduction Biology. DOI: https://doi.org/10.1016/B978-0-12-812351-5.00004-0

© 2018 Elsevier Inc. All rights reserved.

THE 78 KDA GLUCOSE-REGULATED PROTEIN (GRP78)

GRP78 Protein Structure

GRP78 is a member of the evolutionarily conserved heat shock 70 family of proteins.[1] This soluble endoplasmic reticulum (ER)-resident protein consists of two major functional domains: A N-terminal domain with ATPase activity and a COOH-terminal polypeptide binding domain containing a lysine-aspartate-glutamate-leucine (KDEL) ER-retention sequence.[2] The substrate binding domain of GRP78 contains two β sheets each composed of four antiparallel β strands as revealed by X-ray crystallography of the *Escherichia coli* homologue, DnaK. Together, the β sheets form a sandwich containing five helical elements.[3] Polypeptides bind to the channel formed by the β sandwich and are maintained within the binding domain by the helical elements which form a "hinged-lid".[3] The KDEL sequence is situated at the COOH-terminal end of GRP78 and is expressed by many ER chaperones. KDEL receptors on the cytosolic side of the ER membrane interact with GRP78 molecules at the KDEL motif and are thus involved in the retention of GRP78 to the ER lumen.[4] While the conventional role of GRP78 is most notably an ER-resident molecular chaperone,[5] further studies have shown that GRP78 is expressed on the cell surface of different cell types during many pathological conditions, including cancer and cardiovascular disease.[6]

TRANSLOCATION OF GRP78 TO THE CELL SURFACE

As noted in Chapter 2, The Endoplasmic Reticulum Chaperone GRP78 Also Functions as a Cell Surface Signaling Receptor, GRP78 is an ER-resident chaperone localized within the ER lumen by its KDEL retention sequence. However, various KDEL-expressing chaperones such as protein disulfide isomerase (PDI) and calreticulin have been documented on the surface of some cell lines.[7] Previous studies have also demonstrated the presence of GRP78 on the cell surface, as well as in cell culture medium and the circulation.[8] Additionally, cell surface GRP78 (CS-GRP78) contains an intact KDEL sequence suggesting that preferential cleavage of this motif is not a requirement for surface expression.[2] Interestingly, the majority of these studies used cancer cells or cells overexpressing GRP78, suggesting that CS-GRP78 may only occur under conditions where it is overexpressed.[9–11] While CS-GRP78 is well documented, the orientation of GRP78 on the cell surface has not been completely elucidated. Another study has reported that CS-GRP78 is tethered by GPI-anchored proteins to the cell surface.[12] Further, there are no structural differences between cytosolic and CS-GRP78.[13] The mechanism through which GRP78 translocates to the plasma membrane remains unknown since CS-CRP78 is documented to retain its KDEL sequence.[2] There are three current possibilities as to how CS-CRP78 escapes the ER retrieval system. Firstly, during ER stress which can lead to GRP78 upregulation, the KDEL receptors may become saturated allowing excess GRP78 to translocate to the cell surface. Secondly, the down regulation of the KDEL receptor may occur under certain conditions, thereby allowing GRP78 to move to the cell surface.[9–11] Finally, a large body of evidence is suggestive of another secretory or chaperone protein aiding in the translocation of GRP78 to the cell surface.[1] Whether these putative GRP78 binding partners could mask the KDEL sequence, thereby allowing GRP78 to escape ER retrieval, is currently unknown.

REGULATING THE LEVELS OF CELL SURFACE GRP78

Hypoxia,[14] ER stress,[15] and the cochaperone murine tumor cell DnaJ-like protein 1 (MTJ-1)[1] have all been shown to regulate cell surface levels of GRP78. Hypoxia has been identified as a mechanism of increasing surface levels of GRP78 in HT1080 human fibrosarcoma cells.[14] Cell surface levels of GRP78 were increased over fourfold as determined by immunohistochemical analysis.[14] ER stress has also been shown to increase not only ER-resident GRP78, but also cell surface and secreted levels of GRP78.[15] Induction of ER stress with thapsigargin demonstrated a marked increase in CS-CRP78 and secreted GRP78 as measured by a biotinylation/streptavidin pulldown assay and ELISA, respectively.[15] However, the mechanisms by which ER stress induces GRP78 translocation to the cell surface, as well as its release into the media, are not well understood.[15] It is worth noting that translocation of GRP78 to the cell surface is not solely dependent on ER stress. Studies have shown that overexpression of GRP78 is sufficient to induce the movement of GRP78 from the ER to the cell surface.[2] Based on these findings, it is possible that the KDEL receptors on the ER become overwhelmed with an increase in GRP78 levels, thus resulting in incomplete GRP78 retention to the ER.[2] Previous investigations suggest that a mutation in the KDEL region of GRP78 reduces localization to the ER.[9] Further investigation is required to determine whether conditions that reduce the expression of the KDEL receptor alter the levels of CS-CRP78.

In addition to hypoxia and ER stress, further studies have demonstrated that the GRP78 cochaperone murine tumor cell DnaJ-like protein 1 (MTJ-1) also has the ability to regulate cell surface levels of GRP78.[1] MTJ-1 increases the catalytic activity of GRP78 in addition to acting as a cochaperone to aid GRP78 in the translocation and correct folding of nascent polypeptides.[10] GRP78 and MTJ-1 associate together on the cell surface as determined by their coimmunoprecipitation from plasma membrane lysates.[1] It is possible that MTJ-1 can block access to the KDEL sequence of GRP78, thus allowing GRP78 to escape the ER retention by the KDEL receptor. Furthermore, silencing of the MTJ-1 gene by RNA interference blocks GRP78 from reaching the cell surface.[1]

NONCHAPERONE FUNCTIONS OF GRP78

Ca^{2+} Binding Protein

Like most molecular chaperones, GRP78 is a major Ca^{2+} binding protein that aids in maintaining ER Ca^{2+} balance.[11] GRP78, as well as other major Ca^{2+} binding chaperones, does not contain a putative Ca^{2+} binding domain; however Ca^{2+} binds to paired anionic amino acid residues within the chaperone's structure.[16] Although GRP78 contains the greatest number of anionic residues, totaling 111, it contains only 19 paired anionic residues, making its Ca^{2+} binding capacity 1–2 mole Ca^{2+}/mole GRP78. This is considerably less than PDI (23 mole Ca^{2+}/mole PDI) and calreticulin (20 mole Ca^{2+}/mole calreticulin).[16] However, overexpression of GRP78 can increase the Ca^{2+} storage capacity of the ER[11] and attenuate oxidant-induced fluctuations in Ca^{2+} and subsequent apoptosis.[17] Additionally, knock-down of GRP78 using antisense RNA increases intracellular-free Ca^{2+} in response to hydrogen peroxide,[17] thereby supporting the role of GRP78 in maintaining cellular Ca^{2+} homeostasis and preventing Ca^{2+} induced apoptosis.

The binding of Ca^{2+} to GRP78 may also influence its ATP/ADP binding domain.[18] Ca^{2+} binding to GRP78 allows for enhanced dissociation of ADP resulting in more efficient ATP binding.[18] Since the binding of GRP78 to caspase-7 is ATP dependent,[13] Ca^{2+} binding to GRP78 may also have consequences on its cytoprotective properties.[18]

GRP78 AND PATHOLOGY

The majority of studies have reported the presence of CS-CRP78 on cultured cancer cells or cells overexpressing GRP78. Additionally, CS-CRP78 levels correlate with various pathological conditions such as rheumatoid arthritis,[19] atherosclerosis,[6] and many cancers, including prostate cancer and melanoma.[20] Patients affected by these various pathologies were shown to have circulating anti-GRP78 autoantibodies that can bind to CS-CRP78[11] (Fig. 4.1, step 1). Under these pathological conditions, it is likely that the persistent presentation of GRP78 on the cell surface results in an autoimmune response, thereby causing the generation of anti-GRP78 autoantibodies.[22]

Rheumatoid arthritis is characterized by chronic inflammation of the synovial tissue of joints and is the leading form of crippling autoimmune disease.[23] GRP78 has been identified on the cell surface of synovial cells and was demonstrated to be highly antigenic.[19] In the case of inflammation and CS-CRP78 expression, specific T cells may be activated by their cognate antigen and rendered proliferation-competent. The physiological presentation of GRP78 on the cell surface may in itself be an indicator of stress in the immune system. In other words, the immune system manages the expression of GRP78 on the cell surface by generating these autoantibodies, although the exact mechanism remains unclear.[19] One study demonstrated that 63% of rheumatoid arthritis patients tested positive for anti-GRP78 autoantibodies, while none of the healthy volunteers had anti-GRP78 autoantibodies.[19] Furthermore, rat models of arthritis develop anti-GRP78 autoantibodies, however intravenous injection of human GRP78 (97% identical with rat GRP78) into these rats, as a means of neutralizing these autoantibodies, prior to arthritis development completely inhibited arthritis.[23]

A study identifying unique molecular targets on the surface of atherosclerotic lesion-resident endothelial cells demonstrated the presence of CS-CRP78.[6] Using in vivo phage biopanning in the atherosclerotic $ApoE^{-/-}$ mouse model as well as human lesions ex vivo, phage preferentially binding to CS-CRP78 on the endothelium were identified.[6]

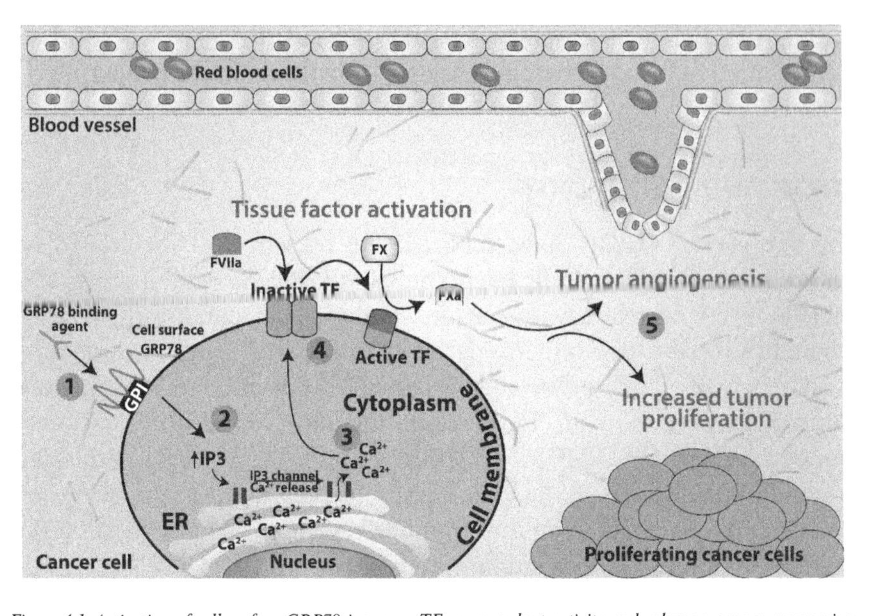

Figure 4.1 Activation of cell surface GRP78 increases TF procoagulant activity and enhances cancer progression. Based on previous and current studies (21, Al-Hashimi et al., 2016 unpublished data), binding of a cellular factor, such as anti-GRP78 autoantibodies or alpha-2-macroglobulin, to the N-terminal domain of cell surface GRP78 (step 1) leads to the activation of an intracellular pathway involving the Gqα11 family of proteins. This results in an increase in cytosolic levels of IP3 molecules (step 2) which can lead to Ca^{2+} efflux from the ER to the cytosol (step 3). Increased cytosolic Ca^{2+} concentrations can alter the plasma membrane leading to the exposure of phosphatidylserine molecules on the cell surface and subsequent induction of TF procoagulant activity (step 4). Consequences of enhanced TF procoagulant activity include increased tumor proliferation potential and angiogenic capacity (step 5). Thus, defining agents that can bind to cell surface GRP78 without activating it but also impair binding of GRP78 ligands may represent a novel approach to treat cancer in patients with elevated levels of anti-GRP78 autoantibodies.

These GRP78 binding phages did interact with the endothelium distal from the atherosclerotic plaques. Moreover, we have demonstrated a significant increase in anti-GRP78 autoantibodies in 25-week-old $ApoE^{-/-}$ mice compared to younger $ApoE^{-/-}$ mice, and that these particular autoantibodies accelerate the development and progression of atherosclerosis (Crane, Al-Hashimi, and Austin, unpublished data).

Finally, a study by Yerushalmi and colleagues suggested that CS-CRP78 may be used as a marker for breast cancer prognosis.[5] Breast cancer cells presenting CS-CRP78 were shown to be significantly more responsive to chemotherapy, in comparison to cells that did not present the protein.[5] This suggests that CS-CRP78 may indicate to the capacity of breast cancer cells to respond to chemotherapy.[5] However, further studies are required to investigate this observation.

ANTI-GRP78 AUTOANTIBODIES AND CANCER

GRP78 is expressed on the cell surface of many different human cancers.[20] Employing a phage biopanning technique, GRP78 was identified as a tumor specific cell surface antigen.[24] Biotinylation of cell surface proteins revealed GRP78 on the cell surface of neuroblastoma cells (SH-SY5Y), lung adenocarcinoma cells (A549), colon adenocarcinoma cells (LoVo), lymphoblastic leukemia-B cells (Sup-B15), ovarian tumor cells,[25] human rabdomiosarcoma cells,[15] NG108-15 cells,[2] and thymic lymphoma cells.[7] Immunofluorescence was also used to demonstrate CS-CRP78 on the prostate adenocarcinoma cell lines 1-LN, PC-3, and DU145 as well as the DM413 melanoma cell line.[26] Moreover, anti-GRP78 autoantibodies have been isolated from the blood of patients with prostate[26] and kidney cancer (Al-Hashimi, Pinthus, and Austin, unpublished data), and their titers correlate with advancing disease and poorer prognosis.[26] Circulating anti-GRP78 autoantibody titers from healthy volunteers average $\sim 7\ \mu g/mL$ while the average in nonmetastatic prostate cancer patients displays titers of over $60\ \mu g/mL$, with some individual patients expressing as high as $250\ \mu g/mL$.[26] Interestingly, the majority of anti-GRP78 autoantibodies target to a specific N-terminal tertiary epitope on GRP78 (Leu^{98}–Leu^{115}).[26] Further studies have demonstrated that these autoantibodies are biologically important since they increase prostate cancer cell proliferation and survival in vitro.[26] While the mechanisms are still not understood, it is clear that during the progression of atherosclerosis and cancer, GRP78 is expressed on the cell surface and that anti-GRP78 autoantibodies play a significant biological function in pathogenesis.

GRP78 AND THE COAGULATION SYSTEM

GRP78, in its capacity as an ER-resident chaperone, can transiently associate with correctly folded proteins, but forms more stable complexes with misfolded or incompletely assembled proteins.[27] This association involves the binding of GRP78 to hydrophobic motifs exposed on unfolded or unassembled polypeptides.[28] Proteins stably bound to GRP78 are subsequently translocated from the ER into the cytosol for proteasome-dependent degradation.[29] GRP78 expression can be induced by conditions of stress and/or disease, such as in the case of cancer, to alleviate accumulation of misfolded or underglycosylated proteins in the ER.[30] Some studies have shown that increased

GRP78 levels may lead to selective changes in the processing and secretion of certain coagulation factors, including the major coagulation protein tissue factor (TF).[31] Thus, this section will focus on investigating the function of GRP78 as a modulator of TF activation, and outline consequences of this phenomenon on disease pathology.

TISSUE FACTOR

The Role of Tissue Factor in Hemostasis

TF is a transmembrane protein that can act as a receptor for blood circulating Factor VII/VIIa (FVII/VIIa).[32] The TF/FVIIa complex is the major initiator of the coagulation cascade in the blood, leading to the formation of a blood clot, as well as activation of platelets.[33] In cases where injury has occurred, clot formation can be initiated via two main pathways—the intrinsic and extrinsic coagulation pathways that merge into a common final pathway leading to fibrin polymerization and cross-linking. A cascade of zymogen activations amplifies the coagulation cascade with the activated form of one clotting factor catalyzing the activation of the next resulting in amplification of the clotting pathway.[34] TF was observed to be constitutively expressed by pericytes and adventitial fibroblasts surrounding blood vessels.[35] The endothelium of the blood vessel physically separates TF from its circulating ligand FVII/FVIIa to prevent inappropriate activation of the clotting cascade.[35] Damage and retraction of the endothelial barrier leads to exposure of extravascular TF and rapid activation of the clotting cascade ensues. Thus, TF is thought to form a hemostatic enclosure that maintains the integrity of blood vessels.[36] In addition, TF is expressed in certain tissues, such as the heart and brain, and provides additional hemostatic protection to these tissues. Small amounts of TF are also present in blood in the form of microparticles, which are small membrane vesicles derived from activated and apoptotic cells.[37] Levels of TF-containing microparticles increase in a variety of diseases, such as sepsis and cancer, and this so-called "blood-borne" lTF may contribute to thrombosis associated with these diseases.[38]

TISSUE FACTOR STRUCTURE

TF is a 263 amino acid transmembrane protein and a member of the type II cytokine receptor family that is responsible for the initiation of the extrinsic pathway of the coagulation cascade.[39] Due to posttranslational N-linked glycosylation, the molecular mass of TF is 45 kDa.[40] TF is encoded by a six exon, five intron 12.4 kb gene located on chromosome 1 of locus 1p22-23.[41] This cell surface glycoprotein—the predominant biological role of TF is extracellular—possesses a single pass transmembrane domain and a short cytoplasmic tail.[39] The N-terminal extracellular domain displays characteristics consistent with other cytokine receptors including a ligand (fVIIa) binding site composed of two fibronectin type III domains and intracellular signaling capacity.[42] In terms of its signaling capacity, TF plays a role in tumor metastasis and angiogenesis which are partially unrelated to coagulation.[43] TF also contains two disulfide bonds in the extracellular domain bridging—Cys^{49}–Cys^{57} and Cys^{186}–Cys^{209}.[42] The transmembrane domain of TF is not specifically required for TF function, however membrane association is. Substitution of the TF transmembrane domain with other lipid domains, such as a phosphatidylinositol anchor, preserves TF function.[44] Although the cytoplasmic tail is essential for TF-FVIIa signaling,[45] it is not required for coagulation[44] or TF encryption.[46]

TISSUE FACTOR AND DISEASE

A variety of hemostatic proteins, including TF, participate in tumor progression by modulating hypercoagulability, angiogenesis, metastasis, and tumor survival. The role of TF in disease progression is an exciting area in cancer research and atherosclerosis. Although the molecular mechanisms are still unknown, TF procoagulant activity regulation is emerging as a major player resulting in protumorigenic gene upregulation and protein synthesis.[47]

Aberrant TF expression is characteristic of a variety of human tumors including breast cancer, pancreatic cancer,[48] lung cancer,[49] colorectal cancer,[21] as well as subsets of glioma[50] and prostate cancer.[51] Furthermore, TF expression often correlates with tumor invasiveness and metastatic potential.[21] Normally, constitutive TF expression is confined to the subendothelium, and only interacts with blood after vascular injury. However, it is apparent that with tumor progression, the regulation of TF expression is lost.[36]

Hypoxia, and the loss of the tumor suppressor gene phosphatase and tensin homolog (PTEN) results in TF upregulation and TF procoagulant activity in malignant gliomas.[52] Additionally, the activation of *KRAS* oncogene and loss of the *p53* tumor suppressor gene controls TF expression in colorectal cancer cells in vitro.[53] Furthermore, the control of TF expression was dependent on mitogen-activated protein kinase and phosphatidylinositol 3-kinase.[53] Evidence from studies published by our group and others have indicated that TF can act in a capacity that is different from its hemostatic function.[47] This is attributable to the ability of TF-mediated intracellular signals to transactivate other growth factor receptors, such as EGFR,[54] interact with integrins,[55] and trigger changes in gene expression profiles.[56] This includes well-identified mediators of angiogenesis and inflammation, including VEGF and IL-8.[56] These properties contribute to the formation of tumor-promoting microenvironment[57] and mediate prometastatic effects of TF through interactions with platelets that blunt innate antitumor responses in the circulation.[58] Indeed, subversion of the TF function can contribute to progression of many pathologies including cancer. Thus, understanding its role and deregulation in these various settings may allow the development of therapies aimed to treat a variety of human disorders.

REGULATION OF TISSUE FACTOR ACTIVITY

Cell surface levels of TF do not directly correlate with TF procoagulant activity, suggesting that TF exists on the cell surface in two distinct classes: inactive (encrypted) and active (deencrypted).[59] TF encryption provides the cell with a mechanism of rapid hemostatic response without transcriptional upregulation of the TF gene. Many agents can significantly increase TF procoagulant activity, including detergents,[60] Ca^{2+} ionophores,[61] and oxidants.[62] TF expression and activity are independently regulated where some agents such as lipoproteins increase de novo TF synthesis, while others including oxidants promote the posttranslational modification of encrypted TF to its active form.[63] For example, the increase in TF procoagulant activity associated with oxLDL treatment drives the transcription of TF mRNA, while treatment with hydrogen peroxide only results in increased TF procoagulant activity.[63]

Since a large portion of expressed TF is expressed on the cell surface, various regulatory pathways are needed to maintain TF in an encrypted state to prevent unprovoked activation of coagulation.[36]

There are several proposed mechanisms regulating cell surface TF encryption including the formation of TF homodimers, the phospholipid microenvironment including phosphatidylserine (PS) exposure, TF compartmentalization in lipid rafts, the endocytosis and degradation of the TF-FVIIa complex, and isomerization of the TF disulfide bond.[64] Although there are several proposed mechanisms of TF post-translational regulation, the mode of TF encryption varies between cell systems. Additionally, the mechanisms through which TF is decrypted are poorly understood and highly controversial.

TISSUE FACTOR DIMERIZATION

The dimerization of exposed TF, or the clustering of TF at the cell surface, has been proposed as a mechanism of TF activation, and that deencryption requires the dissociation of TF homodimers into monomers.[65] Although TF dimers are still capable of binding FVIIa, dimerization can block the binding site of TF via steric hindrance, thus preventing the formation of the coagulation initiation complex (TF-FVIIa) and preventing activation of the coagulation cascade.[65] Studies have shown that an increase in cytosolic Ca^{2+} has been demonstrated to reverse TF dimerization.[65] Furthermore, it has been demonstrated that the extracellular domain of TF dimerizes in a purified in vitro system, and that this does not influence FXa generation, however it did increase FVIIa autoactivation.[66] Additionally, the observation that TF dimerization does not decrease FXa generation is contradictory to the TF dimerization model. Although this suggested model of TF regulation could aid in TF encryption, studies imply that alone, TF dimerization likely does not influence TF procoagulant activity.[66]

ALTERED PHOSPHOLIPID ASYMMETRY

The plasma membrane lipid-bilayer is composed of phospholipids distributed asymmetrically between the inner and outer leaflets. Typically, anionic phospholipids including PS and phosphatidylethanolamine are retained on the inner leaflet and choline-containing phospholipids such as phosphatidylcholine (PC) and sphingomyelin reside on the outer leaflet.[67] The lipid transporters flippase, floppase, and scramblase all perform independent functions resulting in the net inward movement of anionic phospholipids to the inner leaflet in quiescent cells dependent on ATP hydrolysis. Flippase is an aminophospholipid translocase that specifically traffics PS inward.[67] Floppase is somewhat less specific and traffics phospholipids outward.[67] Scramblase is Ca^{2+} dependent and traffics lipids bidirectionally.[68] When cells are perturbed and these proteins become enhanced or inhibited, the PS asymmetry can shift outward. A well-established mechanism of cell surface TF regulation is the composition of the phospholipid microenvironment and the availability of anionic phospholipids, most notably PS, on the outer leaflet in close proximity to TF.[64] Studies suggest that PS interacts with the TF-FVIIa complex aiding in the localization of the complex with FX, resulting in a more rapid conversion of FX to FXa.[69] Additionally, the Ca^{2+} cations bound to the Gla domain of FX and other coagulation factors interact with the anionic charge of PS.[70] Furthermore, blocking PS with annexin V has been demonstrated to attenuate Ca^{2+} ionophore induced TF procoagulant activity.[71]

In addition to Ca^{2+} ionophores, other biologically relevant events including platelet activation by collagen and thrombin can disrupt PS asymmetry.[72] Apoptosis also induces PS asymmetry and TF procoagulant activity.[62] Furthermore, many human diseases associated with increased risks of thrombotic events display increased PS exposure, including cancer, antiphospholipid antibody syndrome, diabetes, viral and bacterial infections, sickle cell anemia, malaria, thalassemia, and preeclampsia.[68] It is evident that PS accelerates coagulation on the cell surface, and there is no doubt that mechanisms resulting in the loss of PS asymmetry are directly linked to increased TF procoagulant activity. However, whether PS functions to increase the association of FX with the TF-FVIIa complex is unknown. Either way, maintenance of phospholipid asymmetry is a rational mechanism by which the cell maintains TF encryption.

TISSUE FACTOR COMPARTMENTALIZATION IN CAVEOLAE

TF compartmentalization within caveolae or lipid rafts provides an additional mechanism by which TF encryption is governed. These caveolae are microdomains on the plasma membrane rich in glycosphingolipids and cholesterol, poor in anionic phospholipids, and contain caveolin-1 expression.[73] Lipid rafts are also microdomains rich in sphingomyelin and cholesterol and can exist as caveolae.[74] Proponents of this theory suggest that this mechanism functions by sequestering TF, thereby increasing self-association and decreasing activity,[75] while also providing an environment nonconducive to TF activation.[73] Evidence supporting this theory is the phenomenon by which TF procoagulant activity is greatly enhanced by the freeze-thawing of cells, which in turn has been demonstrated to disrupt caveolae.[76] This theory suggests that FVIIa and TF antibodies are still capable of binding caveolae-sequestered TF.[76] However the TF-FVIIa complexes formed within these microenvironments are not able to activate FX, suggesting that TF remains encrypted.

TISSUE FACTOR DISULFIDE BOND ISOMERIZATION

A novel controversial mechanism of TF encryption suggests that cryptic TF contains unpaired cysteine thiols on the extracellular domain at Cys^{186} and Cys^{209} and that decryption requires the formation of a Cys^{186}–Cys^{209} disulfide bond.[77] TF contains four cysteine residues on the extracellular domain and the disulfide bond between Cys^{186} and Cys^{209} is allosteric since it bonds cysteines on adjacent strands of a single β-sheet.[78] It is well established that encrypted TF has a much lower affinity for FVIIa than active TF and many have reasoned that this change in affinities is due to a conformational change in TF. The post-translational modification of disulfide bond formation between Cys^{186} and Cys^{209} satisfies this reasoning and has been shown to be required for the activation of coagulation.[69] Further data supports the integral role played by the Cys^{186}–Cys^{209} disulfide bond in maintaining the state of TF encryption.[77] Additionally, adenoviral transduction of human umbilical vein endothelial cells with alanine substitutions at Cys^{186} and Cys^{209} prevented disulfide bond formation, demonstrating impairment in coagulation. However, TF signaling remained intact.[79] Oxidation of the disulfide bond has been demonstrated to alter the quaternary structure of TF and improve substrate binding to TF.[78]

PDI is an oxidoreductase chaperone that functions in the ER to shuffle disulfide bonds during nascent polypeptide folding.[80] Interestingly, colocalization of PDI and TF on the cell surface has been demonstrated using biotinylation assays and immunoprecipitation. PDI has been proposed as the switch responsible for TF disulfide bond oxidation and reduction resulting in TF procoagulant activity or TF encryption and cell signaling respectively.[77] The controversy surrounding this hypothesis is again worth mentioning since other groups have published contradictory results. These studies did not identify cell surface PDI or colocalization of PDI with TF.[81] Additionally, shRNA knock-down of PDI has no effect on TF signaling or procoagulant function.[81] Opponents of this hypothesis call attention to PDI being predominantly located with the ER without a proper membrane attachment site, inferring that cell surface levels of PDI would be in trace amounts at best. However, like GRP78, conditions that allow for its movement from the ER to the cell surface, would potentially enhance its interaction with TF.

REGULATION OF TISSUE FACTOR BY CELL SURFACE GRP78

Despite the central role of TF in blood coagulation and its association with an increased risk of thrombosis, the mechanisms responsible for TF expression and the cellular factors that mediate TF deencryption are poorly understood. Our previous report demonstrated that GRP78 overexpression inhibits TF procoagulant activity in a prothrombotic bladder carcinoma cell line.[82] Stable overexpression of GRP78 attenuates TF-mediated thrombin generation without decreasing cell surface levels of TF.[82] These findings suggest that GRP78 inhibition of TF procoagulant activity is not due to ER retention, but rather an indirect Ca^{2+}-mediated mechanism since the overexpression of GRP78 was also capable of decreasing TF procoagulant activity induced by ionomycin, hydrogen peroxide, and adenoviral infection.[82]

A study by Bhattacharjee and colleagues suggest that CS-CRP78 may attenuate TF procoagulant activity on the cell surface via direct interaction.[83] Further, commercial antibodies targeting the N-terminal domain of CS-CRP78 were thought to abolish this interaction and, thus, increase TF procoagulant activity.[83] One caveat to this study is the fact that experiments were conducted in endothelial cells overexpressing TF which could potentially enhance its interaction with endogenous levels of GRP78. Similar to the previous study,[83] we redundant-investigated the effect of pathological titers of autoantibodies purified from prostate cancer patients (anti-GRP78 autoantibodies) on increased TF procoagulant activity.[60] Anti-GRP78 autoantibodies bind to the N-terminal domain of CS-CRP78, thereby causing ER Ca^{2+} efflux into the cytosol. Increased cytosolic Ca^{2+} levels can increase TF procoagulant activity, which is similar to the effects observed for the Ca^{2+} ionophore, ionomycin.[65] Activation of CS-CRP78 results in Gqα-protein-dependent activation of phospholipase C (PLC) and inositol triphosphate (IP3)-mediated ER Ca^{2+} release[84] (Fig. 4.1, step 2). PLC cleaves phosphatidylinositol 4,5-bisphosphate (PIP2) to produce diacyl glycerol and IP3. Once present in the cytosol, IP3 molecules bind to specific IP3 receptors/Ca^{2+} channels on the ER membrane,[85] resulting in the opening of ER Ca^{2+} channels and increased cytosolic [Ca^{2+}] (Fig. 4.1, step 3). An increase in cytosolic Ca^{2+} levels can lead to activation of TF[60] (Fig. 4.1, step 4). Thus, the binding and activation of the anti-GRP78 autoantibodies to the N-terminal domain of CS-CRP78 leads to increased TF procoagulant activity via a Ca^{2+}-mediated mechanism. Although both GRP78 and TF are present on the cell surface of cancer cells, there was no evidence of a stable interaction at the cell surface.[60] Consistent with these findings, the interaction of human recombinant GRP78 with TF using a purified system showed no significant interaction.[60] Taken together, these findings suggest that CS-CRP78 can modulate TF procoagulant activity indirectly by regulating levels of cytosolic Ca^{2+}.

It is worth noting that other circulating blood factors, such as alpha-2-macroglobulin, can bind to CS-CRP78.[84] Binding of alpha-2-macroglobulin to CS-GRP78, on tumor cells, causes its activation and results in activation of a downstream pro-proliferative and antiapoptotic signaling cascades.[86] Our studies indicate that activation of CS-CRP78 by alpha-2-macroglobulin leads to upregulation of TF procoagulant activity (Al-Hashimi and Austin, unpublished data). This further highlights the role of CS-CRP78 in activating TF procoagulant activity via binding to factors available in the blood. Thus, defining binding partners of CS-CRP78 that can elicit the intracellular response described previously will allow for a better understanding of how GRP78 regulates TF procoagulant activity.

IMPLICATION OF TF ACTIVATION VIA CELL SURFACE GRP78 IN CANCER

GRP78 is localized to the ER under normal conditions. However, the translocation of GRP78 from the ER to the cell surface can allow it to be recognized by the immune system, thus causing an immune-response and generation of autoantibodies against specific epitopes.[87] We have shown previously that anti-GRP78 autoantibodies isolated from the blood of prostate cancer patients can increase TF procoagulant activity, thereby leading to a condition termed hypercoagulability. Previous studies have shown that hypercoagulability is common in cancer patients resulting in an increased risk of venous thromboembolism (VTE), pulmonary embolism, disseminated intravascular coagulopathy, and hemorrhage.[88] Moreover, this hypercoagulable state can contribute to tumorigenesis by supporting angiogenesis—TF-mediated angiogenesis occurs indirectly through clotting[38] or through the release of proangiogenic factors[89] as well as directly through clotting-independent mechanisms[90] (Fig. 4.1, step 5). Using a breast cancer cell model, the observed hypercoagulability is dependent on TF—clot formation was FVIIa dependent and attenuated by anti-TF antibodies.[91] Additionally, the upregulation of TF is an independent predictor of VTE in ovarian cancer[92] and is associated with VTE in pancreatic cancer.[93] Furthermore, the upregulation of TF has also been demonstrated to lead to the systemic release of TF positive microparticles.[53] TF positive microparticle levels can predict deep vein thrombosis DVT and correlate with D-dimer levels—which strongly predict VTE in cancer patients.[94]

Another consequence of enhanced TF activation via CS-CRP78 is an overall increase in the metastatic potential of cells. Metastasis is the process through which tumor cells infiltrate the bloodstream or lymphatic vessels and migrate to distant sites to establish secondary tumors. This process is vastly dependent on the coagulation cascade, most notably on thrombin generation.[95] TF has been demonstrated to play a supporting role through fibrinogen- and platelet-dependent restriction of natural killer cells—mediate micrometastases clearance.[58] One study suggested that platelets have the capacity to present GRP78 on the cell surface; a further blockade of CS-CRP78 resulted in a substantial increase in platelet deposition and TF procoagulant activity.[96]

Many studies have linked TF-FVIIa signaling to cell motility. Cleaving of the TF extracellular domain was demonstrated to result in the interaction of the TF cytoplasmic domain with actin-binding protein 280 which is involved in cell adhesion and migration in vitro.[97] Moreover, the protease activated receptors (PARs) mediators of coagulation factor signaling were implicated in conjunction with tyrosine kinase receptors in a study linking TF-FVIIa signaling and migration stimulated by platelet derived growth factor-BB.[98] PAR-1 is highly expressed and is activated by thrombin in invasive breast cancers, but is absent in healthy breast tissue or noninvasive cancers.[95] A PAR-2 mechanism is involved in the migration of breast cancer cells in vitro independent of TF-FVIIa signaling.[99] The cytoplasmic domain of TF proves to be critical in the migration of porcine aortic endothelial[100] and human bladder carcinoma cells.[45] An additional mechanism through which TF promotes tumor cell migration and survival is through protection against immune recognition and cytotoxicity. TF expressing colon cancer cells demonstrated an increased invasion rate from peripheral blood monocytes dependent on TF.[101] This could lead to increased cell survival and cancer progression, however, the exact mechanism is not known. Collectively, the role of CS-CRP78 in cancer can be viewed as a unique signaling factor that can mediate disease progression via TF. Based on published studies, the mechanism by which CS-CRP78 increases TF signaling and its consequence on cancer progression is summarized in Fig. 4.1.

CONCLUSION

The translocation of GRP78 from the ER lumen to the cell surface under pathological conditions can impact disease progression. In addition to its function as a procoagulant protein, studies have identified activated TF as a significant contributor to cancer progression. A growing body of evidence has shown that activation of CS-CRP78 via binding to anti-GRP78 autoantibodies or alpha-2-macroglobulin results in enhanced TF procoagulant activity, therefore, contributing to cancer progression. A critical question still remains as to whether TF signaling is more relevant to disease progression than its procoagulant activity. Moreover, further investigations are required to determine the immune system profile, including inflammation status that can lead to generation of anti-GRP78 autoantibodies. Such inquiries will assist designing new theranostics approaches to exploit how targeting of CS-CRP78 will reduce disease progression and detection of prostate cancer at an earlier stage. Understanding such mechanisms can be viewed as a platform for the development of new therapies that can target the rate of disease progression, and thus may improve the efficacy of current treatments.

REFERENCES

1. Misra UK, Gonzalez-Gronow M, Gawdi G, Pizzo SV. The role of MTJ-1 in cell surface translocation of GRP78, a receptor for alpha 2-macroglobulin-dependent signaling. *J Immunol* 2005;**174**:2092–7.
2. Xiao G, Chung TF, Pyun HY, Fine RE, Johnson RJ. KDEL proteins are found on the surface of NG108-15 cells. *Brain Res Mol Brain Res* 1999;**72**:121–8.
3. Gething MJ, Sambrook J. Protein folding in the cell. *Nature* 1992;**355**:33–45.
4. Munro S, Pelham HR. A C-terminal signal prevents secretion of luminal ER proteins. *Cell* 1987;**48**:899–907.
5. Ron D, Walter P. Signal integration in the endoplasmic reticulum unfolded protein response. *Nat Rev Mol Cell Biol* 2007;**8**(7):519–29.
6. Liu C, Bhattacharjee G, Boisvert W, Dilley R, Edgington T. In vivo interrogation of the molecular display of atherosclerotic lesion surfaces. *Am J Pathol* 2003;**163**:1859–71.
7. Wiest DL, Bhandoola A, Punt J, Kreibich G, McKean D, Singer A. Incomplete endoplasmic reticulum (ER) retention in immature thymocytes as revealed by surface expression of "ER-resident" molecular chaperones. *Proc Natl Acad Sci USA* 1997;**94**:1884–9.
8. Bugge TH, Xiao Q, Kombrinck KW, Flick MJ, Holmback K, Danton MJ, et al. Fatal embryonic bleeding events in mice lacking tissue factor, the cell-associated initiator of blood coagulation. *Proc Natl Acad Sci USA* 1996;**93**:6258–63.
9. Mimura N, Yuasa S, Soma M, Jin H, Kimura K, Goto S, et al. Altered quality control in the endoplasmic reticulum causes cortical dysplasia in knock-in mice expressing a mutant BiP. *Mol Cell Biol* 2008;**28**:293–301.
10. Chevalier M, Rhee H, Elguindi EC, Blond SY. Interaction of murine BiP/GRP78 with the DnaJ homologue MTJ1. *J Biol Chem* 2000;**275**:19620–7.
11. Lievremont JP, Rizzuto R, Hendershot L, Meldolesi J. BiP, a major chaperone protein of the endoplasmic reticulum lumen, plays a direct and important role in the storage of the rapidly exchanging pool of Ca^{2+}. *J Biol Chem* 1997;**272**:30873–9.
12. Tsai YL, Zhang Y, Tseng CC, Stanciauskas R, Pinaud F, Lee AS. Characterization and mechanism of stress-induced translocation of 78-kilodalton glucose-regulated protein (GRP78) to the cell surface. *J Biol Chem* 2015;**290**:8049–64.
13. Reddy RK, Mao C, Baumeister P, Austin RC, Kaufman RJ, Lee AS. Endoplasmic reticulum chaperone protein GRP78 protects cells from apoptosis induced by topoisomerase inhibitors: role of ATP binding site in suppression of caspase-7 activation. *J Biol Chem* 2003;**278**:20915–24.
14. Davidson DJ, Haskell C, Majest S, Kherzai A, Egan DA, Walter KA, et al. Kringle 5 of human plasminogen induces apoptosis of endothelial and tumor cells through surface-expressed glucose-regulated protein 78. *Cancer Res* 2005;**65**:4663–72.
15. Delpino A, Castelli M. The 78 kDa glucose-regulated protein (GRP78/BIP) is expressed on the cell membrane, is released into cell culture medium and is also present in human peripheral circulation. *Biosci Rep* 2002;**22**:407–20.
16. Lucero HA, Lebeche D, Kaminer B. ERcalcistorin/protein disulfide isomerase (PDI). Sequence determination and expression of a cDNA clone encoding a calcium storage protein with PDI activity from endoplasmic reticulum of the sea urchin egg. *J Biol Chem* 1994;**269**:23112–19.
17. Aoki M, Tamatani M, Taniguchi M, Yamaguchi A, Bando Y, Kasai K, et al. Hypothermic treatment restores glucose regulated protein 78 (GRP78) expression in ischemic brain. *Brain Res Mol Brain Res* 2001;**95**:117–28.

18. Lamb HK, Mee C, Xu W, Liu L, Blond S, Cooper A, et al. The affinity of a major Ca^{2+} binding site on GRP78 is differentially enhanced by ADP and ATP. *J Biol Chem* 2006;**281**:8796–805.

19. Blass S, Union A, Raymackers J, Schumann F, Ungethum U, Muller-Steinbach S, et al. The stress protein BiP is overexpressed and is a major B and T cell target in rheumatoid arthritis. *Arthritis Rheum* 2001;**44**:761–71.

20. Mintz PJ, Kim J, Do KA, Wang X, Zinner RG, Cristofanilli M, et al. Fingerprinting the circulating repertoire of antibodies from cancer patients. *Nat Biotechnol* 2003;**21**:57–63.

21. Lykke J, Nielsen HJ. The role of tissue factor in colorectal cancer. *Eur J Surg Oncol* 2003;**29**:417–22.

22. Lee GY, Kim SK, Byun Y. Glucosylated heparin derivatives as non-toxic anti-cancer drugs. *J Control Release* 2007;**123**:46–55.

23. Corrigall VM, Bodman-Smith MD, Fife MS, Canas B, Myers LK, Wooley P, et al. The human endoplasmic reticulum molecular chaperone BiP is an autoantigen for rheumatoid arthritis and prevents the induction of experimental arthritis. *J Immunol* 2001;**166**:1492–8.

24. Arap MA, Lahdenranta J, Mintz PJ, Hajitou A, Sarkis AS, Arap W, et al. Cell surface expression of the stress response chaperone GRP78 enables tumor targeting by circulating ligands. *Cancer Cell* 2004;**6**:275–84.

25. Shin BK, Wang H, Yim AM, Le Naour F, Brichory F, Jang JH, et al. Global profiling of the cell surface proteome of cancer cells uncovers an abundance of proteins with chaperone function. *J Biol Chem* 2003;**278**:7607–16.

26. Gonzalez-Gronow M, Cuchacovich M, Llanos C, Urzua C, Gawdi G, Pizzo SV. Prostate cancer cell proliferation in vitro is modulated by antibodies against glucose-regulated protein 78 isolated from patient serum. *Cancer Res* 2006;**66**:11424–31.

27. Klausner RD, Sitia R. Protein degradation in the endoplasmic reticulum. *Cell* 1990;**62**:611–14.

28. Knarr G, Modrow S, Todd A, Gething MJ, Buchner J. BiP-binding sequences in HIVgp160. Implications for the binding specificity of bip. *J Biol Chem* 1999;**274**:29850–7.

29. Meerovitch K, Wing S, Goltzman D. Proparathyroid hormone-related protein is associated with the chaperone protein BiP and undergoes proteasome-mediated degradation. *J Biol Chem* 1998;**273**:21025–30.

30. Lee AS. The glucose-regulated proteins: stress induction and clinical applications. *Trends Biochem Sci* 2001;**26**:504–10.

31. Dorner AJ, Kaufman RJ. Analysis of synthesis, processing, and secretion of proteins expressed in mammalian cells. *Methods Enzymol* 1990;**185**:577–96.

32. Mackman N. Role of tissue factor in hemostasis, thrombosis, and vascular development. *Arterioscler Thromb Vasc Biol* 2004;**24**:1015–22.

33. Bach RR. Initiation of coagulation by tissue factor. *CRC Crit Rev Biochem* 1988;**23**:339–68.

34. Bergum PW, Cruikshank A, Maki SL, Kelly CR, Ruf W, Vlasuk GP. Role of zymogen and activated factor X as scaffolds for the inhibition of the blood coagulation factor VIIa-tissue factor complex by recombinant nematode anticoagulant protein c2. *J Biol Chem* 2001;**276**:10063–71.

35. Flossel C, Luther T, Muller M, Albrecht S, Kasper M. Immunohistochemical detection of tissue factor (TF) on paraffin sections of routinely fixed human tissue. *Histochemistry* 1994;**101**:449–53.

36. Drake TA, Morrissey JH, Edgington TS. Selective cellular expression of tissue factor in human tissues. Implications for disorders of hemostasis and thrombosis. *Am J Pathol* 1989;**134**:1087–97.

37. Del Conde I, Shrimpton CN, Thiagarajan P, Lopez JA. Tissue-factor-bearing microvesicles arise from lipid rafts and fuse with activated platelets to initiate coagulation. *Blood* 2005;**106**:1604–11.

38. Milsom C, Rak J. Tissue factor and cancer. *Pathophysiol Haemost Thromb* 2008;**36**:160–76.

39. Martin DM, Boys CW, Ruf W. Tissue factor: molecular recognition and cofactor function. *FASEB J* 1995;**9**:852–9.

40. Spicer EK, Horton R, Bloem L, Bach R, Williams KR, Guha A, et al. Isolation of cDNA clones coding for human tissue factor: primary structure of the protein and cDNA. *Proc Natl Acad Sci USA* 1987;**84**:5148–52.

41. Mackman N, Morrissey JH, Fowler B, Edgington TS. Complete sequence of the human tissue factor gene, a highly regulated cellular receptor that initiates the coagulation protease cascade. *Biochemistry* 1989;**28**:1755–62.

42. Harlos K, Martin DM, O'Brien DP, Jones EY, Stuart DI, Polikarpov I, et al. Crystal structure of the extracellular region of human tissue factor. *Nature* 1994;**370**:662–6.

43. Bromberg ME, Konigsberg WH, Madison JF, Pawashe A, Garen A. Tissue factor promotes melanoma metastasis by a pathway independent of blood coagulation. *Proc Natl Acad Sci USA* 1995;**92**:8205–9.

44. Paborsky LR, Caras IW, Fisher KL, Gorman CM. Lipid association, but not the transmembrane domain, is required for tissue factor activity. Substitution of the transmembrane domain with a phosphatidylinositol anchor. *J Biol Chem* 1991;**266**:21911–16.

45. Ott I, Weigand B, Michl R, Seitz I, Sabbari-Erfani N, Neumann FJ, et al. Tissue factor cytoplasmic domain stimulates migration by activation of the GTPase Rac1 and the mitogen-activated protein kinase p38. *Circulation* 2005;**111**:349–55.

46. Wolberg AS, Kon RH, Monroe DM, Ezban M, Roberts HR, Hoffman M. Deencryption of cellular tissue factor is independent of its cytoplasmic domain. *Biochem Biophys Res Commun* 2000;**272**:332–6.

47. Rak J, Milsom C, May L, Klement P, Yu J. Tissue factor in cancer and angiogenesis: the molecular link between genetic tumor progression, tumor neovascularization, and cancer coagulopathy. *Semin Thromb Hemost* 2006;**32**:54–70.

48. Vrana JA, Stang MT, Grande JP, Getz MJ. Expression of tissue factor in tumor stroma correlates with progression to invasive human breast cancer: paracrine regulation by carcinoma cell-derived members of the transforming growth factor beta family. *Cancer Res* 1996;**56**:5063–70.

49. Koomagi R, Volm M. Tissue-factor expression in human non-small-cell lung carcinoma measured by immunohistochemistry: correlation between tissue factor and angiogenesis. *Int J Cancer* 1998;**79**:19–22.

50. Hamada K, Kuratsu J, Saitoh Y, Takeshima H, Nishi T, Ushio Y. Expression of tissue factor correlates with grade of malignancy in human glioma. *Cancer* 1996;**77**:1877–83.

51. Abdulkadir SA, Carvalhal GF, Kaleem Z, Kisiel W, Humphrey PA, Catalona WJ, et al. Tissue factor expression and angiogenesis in human prostate carcinoma. *Hum Pathol* 2000;**31**:443–7.

52. Yu JL, May L, Klement P, Weitz JI, Rak J. Oncogenes as regulators of tissue factor expression in cancer: implications for tumor angiogenesis and anti-cancer therapy. *Semin Thromb Hemost* 2004;**30**:21–30.

53. Yu JL, May L, Lhotak V, Shahrzad S, Shirasawa S, Weitz JI, et al. Oncogenic events regulate tissue factor expression in colorectal cancer cells: implications for tumor progression and angiogenesis. *Blood* 2005;**105**:1734–41.

54. Prydz H, Pettersen KS. Synthesis of thromboplastin (tissue factor) by endothelial cells. *Haemostasis* 1988;**18**:215–23.
55. Versteeg HH, Ruf W. Emerging insights in tissue factor-dependent signaling events. *Semin Thromb Hemost* 2006;**32**:24–32.
56. Camerer E, Gjernes E, Wiiger M, Pringle S, Prydz H. Binding of factor VIIa to tissue factor on keratinocytes induces gene expression. *J Biol Chem* 2000;**275**:6580–5.
57. Magnussen S, Hadler-Olsen E, Latysheva N, Pirila E, Steigen SE, Hanes R, et al. Tumour microenvironments induce expression of urokinase plasminogen activator receptor (uPAR) and concomitant activation of gelatinolytic enzymes. *PLoS One* 2014;**9**:e105929.
58. Palumbo JS, Degen JL. Mechanisms linking tumor cell-associated procoagulant function to tumor metastasis. *Thromb Res* 2007;**120**(Suppl. 2):S22–28.
59. Bach R, Rifkin DB. Expression of tissue factor procoagulant activity: regulation by cytosolic calcium. *Proc Natl Acad Sci USA* 1990;**87**:6995–9.
60. Al-Hashimi AA, Caldwell J, Gonzalez-Gronow M, Pizzo SV, Aboumrad D, Pozza L, et al. Binding of anti-GRP78 autoantibodies to cell surface GRP78 increases tissue factor procoagulant activity via the release of calcium from endoplasmic reticulum stores. *J Biol Chem* 2010;**285**:28912–23.
61. Caldwell JA, Dickhout JG, Al-Hashimi AA, Austin RC. Development of a continuous assay for the measurement of tissue factor procoagulant activity on intact cells. *Lab Invest* 2010;**90**:953–62.
62. Greeno EW, Bach RR, Moldow CF. Apoptosis is associated with increased cell surface tissue factor procoagulant activity. *Lab Invest* 1996;**75**:281–9.
63. Penn MS, Patel CV, Cui MZ, DiCorleto PE, Chisolm GM. LDL increases inactive tissue factor on vascular smooth muscle cell surfaces: hydrogen peroxide activates latent cell surface tissue factor. *Circulation* 1999;**99**:1753–9.
64. Kunzelmann-Marche C, Satta N, Toti F, Zhang Y, Nawroth PP, Morrissey JH, et al. The influence exerted by a restricted phospholipid microenvironment on the expression of tissue factor activity at the cell plasma membrane surface. *Thromb Haemost* 2000;**83**:282–9.
65. Bach RR, Moldow CF. Mechanism of tissue factor activation on HL-60 cells. *Blood* 1997;**89**:3270–6.
66. Donate F, Kelly CR, Ruf W, Edgington TS. Dimerization of tissue factor supports solution-phase autoactivation of factor VII without influencing proteolytic activation of factor X. *Biochemistry* 2000;**39**:11467–76.
67. Diaz C, Schroit AJ. Role of translocases in the generation of phosphatidylserine asymmetry. *J Membr Biol* 1996;**151**:1–9.
68. Zwaal RF, Comfurius P, Bevers EM. Surface exposure of phosphatidylserine in pathological cells. *Cell Mol Life Sci* 2005;**62**:971–88.
69. Ruf W, Disse J, Carneiro-Lobo TC, Yokota N, Schaffner F. Tissue factor and cell signalling in cancer progression and thrombosis. *J Thromb Haemost* 2011;**9**(Suppl. 1):306–15.
70. Morrissey JH. Tissue factor: an enzyme cofactor and a true receptor. *Thromb Haemost* 2001;**86**:66–74.
71. Wolberg AS, Monroe DM, Roberts HR, Hoffman MR. Tissue factor de-encryption: ionophore treatment induces changes in tissue factor activity by phosphatidylserine-dependent and -independent mechanisms. *Blood Coagul Fibrinolysis* 1999;**10**:201–10.
72. Zwaal RF, Schroit AJ. Pathophysiologic implications of membrane phospholipid asymmetry in blood cells. *Blood* 1997;**89**:1121–32.

73. Mulder AB, Smit JW, Bom VJ, Blom NR, Halie MR, van der Meer J. Association of endothelial tissue factor and thrombomodulin with caveolae. *Blood* 1996;**88**:3667–70.

74. Lupu C, Hu X, Lupu F. Caveolin-1 enhances tissue factor pathway inhibitor exposure and function on the cell surface. *J Biol Chem* 2005;**280**:22308–17.

75. Bach RR. Tissue factor encryption. *Arterioscler Thromb Vasc Biol* 2006;**26**:456–61.

76. Giesen PL, Nemerson Y. Tissue factor on the loose. *Semin Thromb Hemost* 2000;**26**:379–84.

77. Ahamed J, Versteeg HH, Kerver M, Chen VM, Mueller BM, Hogg PJ, et al. Disulfide isomerization switches tissue factor from coagulation to cell signaling. *Proc Natl Acad Sci USA* 2006;**103**:13932–7.

78. Chen VM, Ahamed J, Versteeg HH, Berndt MC, Ruf W, Hogg PJ. Evidence for activation of tissue factor by an allosteric disulfide bond. *Biochemistry* 2006;**45**:12020–8.

79. Rehemtulla A, Ruf W, Edgington TS. The integrity of the cysteine 186-cysteine 209 bond of the second disulfide loop of tissue factor is required for binding of factor VII. *J Biol Chem* 1991;**266**:10294–9.

80. Freedman RB. Post-translational modification and folding of secreted proteins. *Biochem Soc Trans* 1989;**17**:331–5.

81. Pendurthi UR, Ghosh S, Mandal SK, Rao LV. Tissue factor activation: is disulfide bond switching a regulatory mechanism? *Blood* 2007;**110**:3900–8.

82. Watson LM, Chan AK, Berry LR, Li J, Sood SK, Dickhout JG, et al. Overexpression of the 78-kDa glucose-regulated protein/immunoglobulin-binding protein (GRP78/BiP) inhibits tissue factor procoagulant activity. *J Biol Chem* 2003;**278**:17438–47.

83. Bhattacharjee G, Ahamed J, Pedersen B, El-Sheikh A, Mackman N, Ruf W, et al. Regulation of tissue factor--mediated initiation of the coagulation cascade by cell surface grp78. *Arterioscler Thromb Vasc Biol* 2005;**25**:1737–43.

84. Gonzalez-Gronow M, Selim MA, Papalas J, Pizzo SV. GRP78: a multifunctional receptor on the cell surface. *Antioxid Redox Signal* 2009;**11**:2299–306.

85. van Rossum DB, Patterson RL. PKC and PLA2: probing the complexities of the calcium network. *Cell Calcium* 2009;**45**:535–45.

86. Misra UK, Gonzalez-Gronow M, Gawdi G, Hart JP, Johnson CE, Pizzo SV. The role of Grp 78 in alpha 2-macroglobulin-induced signal transduction. Evidence from RNA interference that the low density lipoprotein receptor-related protein is associated with, but not necessary for, GRP 78-mediated signal transduction. *J Biol Chem* 2002;**277**:42082–7.

87. Binder RJ. Hsp receptors: the cases of identity and mistaken identity. *Curr Opin Mol Ther* 2009;**11**:62–71.

88. Prandoni P. Cancer and thromboembolic disease: how important is the risk of thrombosis? *Cancer Treat Rev* 2002;**28**:133–6.

89. Zhang Y, Deng Y, Luther T, Muller M, Ziegler R, Waldherr R, et al. Tissue factor controls the balance of angiogenic and antiangiogenic properties of tumor cells in mice. *J Clin Invest* 1994;**94**:1320–7.

90. Hembrough TA, Swartz GM, Papathanassiu A, Vlasuk GP, Rote WE, Green SJ, et al. Tissue factor/factor VIIa inhibitors block angiogenesis and tumor growth through a nonhemostatic mechanism. *Cancer Res* 2003;**63**:2997–3000.

91. Hu T, Bach RR, Horton R, Konigsberg WH, Todd MB. Procoagulant activity in cancer cells is dependent on tissue factor expression. *Oncol Res* 1994;**6**:321–7.

92. Shao YP, Cheng XD, Wan XY. Expression and clinical significance of GRIM-19 in epithelial ovarian carcinoma. *Zhonghua Fu Chan Ke Za Zhi* 2012;**47**:751–5.

93. Khorana AA, Ahrendt SA, Ryan CK, Francis CW, Hruban RH, Hu YC, et al. Tissue factor expression, angiogenesis, and thrombosis in pancreatic cancer. *Clin Cancer Res* 2007;**13**:2870–5.

94. Al-Nedawi K, Meehan B, Rak J. Microvesicles: messengers and mediators of tumor progression. *Cell Cycle* 2009;**8**:2014–18.

95. Booden MA, Eckert LB, Der CJ, Trejo J. Persistent signaling by dysregulated thrombin receptor trafficking promotes breast carcinoma cell invasion. *Mol Cell Biol* 2004;**24**:1990–9.

96. Molins B, Pena E, Padro T, Casani L, Mendieta C, Badimon L. Glucose-regulated protein 78 and platelet deposition: effect of rosuvastatin. *Arterioscler Thromb Vasc Biol* 2010;**30**:1246–52.

97. Ott I. Soluble tissue factor emerges from inflammation. *Circ Res* 2005;**96**:1217–18.

98. Siegbahn A. Cellular consequences upon factor VIIa binding to tissue factor. *Haemostasis* 2000;**30**(Suppl. 2):41–7.

99. Jiang X, Bailly MA, Panetti TS, Cappello M, Konigsberg WH, Bromberg ME. Formation of tissue factor-factor VIIa-factor Xa complex promotes cellular signaling and migration of human breast cancer cells. *J Thromb Haemost* 2004;**2**:93–101.

100. Siegbahn A, Johnell M, Rorsman C, Ezban M, Heldin CH, Ronnstrand L. Binding of factor VIIa to tissue factor on human fibroblasts leads to activation of phospholipase C and enhanced PDGF-BB-stimulated chemotaxis. *Blood* 2000;**96**:3452–8.

101. Li C, Colman LM, Collier ME, Dyer CE, Greenman J, Ettelaie C. Tumour-expressed tissue factor inhibits cellular cytotoxicity. *Cancer Immunol Immunother* 2006;**55**:1301–8.

Novel Cell Surface Targets for the Plasminogen Activating System in the Brain: Implications for Human Disease

Mario Gonzalez-Gronow and Salvatore V. Pizzo
Duke University Medical Center, Durham, NC, United States

INTRODUCTION

Activation of plasminogen (Pg) is important for regulating cell biology in a number of settings, including the central nervous system (CNS), where unique cell surface protein complexes serve to bind and activate Pg with a number of downstream consequences. One of such complexes involves the voltage-dependent anion channel 1 (VDAC1) in association with the Glucose Regulated Protein 78,000 (GRP78), tissue-type plasminogen activator (t-PA), and Pg. In this chapter, we will discuss the relevance of VDAC1 in association with these three proteins in mechanisms that control VDAC1 activity under normal and pathological conditions, both at the mitochondrial and plasma membrane levels.

Cell Surface GRP78, a New Paradigm in Signal Transduction Biology. DOI: https://doi.org/10.1016/B978-0-12-812351-5.00005-2
© 2018 Elsevier Inc. All rights reserved.

THE VOLTAGE-DEPENDENT ANION CHANNEL 1

VDAC1, is a small 30–35 kDa protein, originally discovered in the outer membrane of mitochondria where it functions as part of the mitochondrial permeability transition pore as a voltage-gated pore, which is important for passive diffusion of substances through the membrane.[1] Three isoforms of VDAC (VDAC 1–3) have been found in humans and in different animal species.[2] VDAC1 is also found in the plasma membrane of a large number of tissues.[3] Studies on rats show that VDAC1 is densely localized in regions of the brain, including the hippocampus, hypothalamus, and the cerebellum.[4]

FUNCTIONS OF VOLTAGE-DEPENDENT ANION CHANNEL

In the mitochondria, VDAC1 transports ions, Ca^{2+}, cholesterol, and metabolites across the outer mitochondrial membrane where it participates in the release of mitochondrial proapoptotic proteins to the cytosol after interaction with apoptosis regulatory proteins.[5] On the cell surface, VDAC is a major player in the maintenance of redox homeostasis in normal cells and also participates in the promotion of anion efflux in apoptotic cells.[6] These functions make VDAC1 a point of convergence for a large variety of cell survival and death signals.[5] A key protein in these mechanisms is hexokinase-I (HK-I), which acts as a gatekeeper by maintaining a delicate balance between the opening and the closing of VDAC.[7] The N-terminal domain of VDAC1 controls the release of cytochrome c, apoptosis, and the regulation of apoptosis by antiapoptotic proteins such as HK-I and the B-cell lymphoma 2 protein (Bcl-2).[8] HK-II competes with Bcl-2 family proteins for binding to VDAC1 and its binding to VDAC1 is regulated by protein kinases, notably glycogen synthase kinase-3beta (GSK-3β) and protein kinase B (Akt), influencing the balance of pro- and antiapoptotic proteins that control outer membrane permeabilization.[9] The Bcl-2 family of proteins regulates mitochondrial permeability.[10] Bax and Bak are associated with the opening of VDAC, whereas the antiapoptotic Bcl-xL closes VDAC by binding to it. Bax and Bak allow cytochrome c to pass through VDAC, but the passage is prevented by Bcl-xL.

The activation of Akt enhances HK-II binding to the mitochondria[11] after phosphorylation of HK-II.[12] Akt also affects the binding of HK to mitochondria by negatively regulating the activity of GSK-3β and its phosphorylation of VDAC on threonine 51 which results in the detachment of HK-II from VDAC.[13] The detachment of HK-II facilitates binding of the proapoptotic protein bcl-2-like protein 4 (Bax) to the mitochondria.[14] Furthermore, activation of GSK-3β phosphorylates Bax directly, promoting its translocation from the cytosol to the mitochondria.[15] Therefore, GSK-3β activation may have the double impact of making mitochondria more vulnerable to Bax by detaching hexokinase and linking the antiapoptotic Bx binding partner Bcl-X_L and then promoting the translocation of Bax.[13] In the brain, these mechanisms are linked to VDAC1, which is the predominant isoform expressed.[16]

Overexpression of VDAC1 over VDAC2 increases cytochrome c release threefold[17] and modulates the type of hexokinase isoform binding on brain mitochondria, suggesting that different ratios of VDAC1 and VDAC2 expression might implicate separate mechanisms in neuronal cell death.[18] VDAC1 at the plasma membrane also participates in mechanisms of hypothermic neuroprotection and improves neurological recovery of animals exposed to focal brain hypoxic-ischemic injury.[19]

THE PLASMINOGEN ACTIVATOR SYSTEM IN THE BRAIN

The Pg activator system plays an important role in the CNS, including processes of neuronal migration, neurite outgrowth and neuronal plasticity.[20,21] The Pg activator, t-PA, may also participate in several neuropathological conditions, such as cerebral ischemia, Alzheimer's disease (AD), and multiple sclerosis.[22] Both Pg and t-PA are expressed in neurons, astrocytes, and microglia and the activity of plasmin (Pm) is tightly regulated[23] by α_2-antiplasmin, the main physiological inhibitor of Pm, which is expressed in the hippocampus, cortex, and cerebellum, where it participates in the dendritic growth of neurons independently of its role as a Pm inhibitor.[24] The activity of t-PA in the CNS is also tightly controlled by the inhibitor plasminogen activator inhibitor-1 (PAI-1), a member of the serpin gene family.[25] If left uncontrolled, any excess of endogenous or exogenous t-PA inhibits dendritic growth in vivo and in vitro via mechanisms that involve the activation of protein kinase $C\gamma$ and phosphorylation of microtubule-associated protein 2.[26] Furthermore, the excess t-PA/Pm impairs

Purkinje neuron synaptogenesis because Pm, bound to VDAC1 via its Kringle 5 (K5), promotes several pathological processes including enlargements and rounding of mitochondria, a reduction of mitochondrial membrane potential, and damage to plasma membranes.[27] Similarly, a fall in the PAI-1 expression, which controls t-PA activity, triggers apoptosis in neuronal cells via a reduction in the expression of Bcl-2 and Bcl-XL and an increase in Bcl-XS and Bax mRNAs. These changes in balance between the anti- and proapoptotic Bcl-2 family proteins cause caspase-3 activation followed by release of cytochrome c from mitochondria.[28] Furthermore, excess plasminogen K5 bound to VDAC promotes phosphorylation of VDAC1 via the Akt-GSK-3β pathway and cell surface translocation of VDAC1, contributing to apoptosis via a dual positive feedback loop involving VDAC1-Akt-GSK-3β-VDAC1, where VDAC1 transmits K5-triggering signals that regulate its own protein level and cell apoptosis.[27]

GRP78 AND VDAC IN THE BRAIN

The VDAC family of proteins are the most abundant proteins of the outer mitochondrial membrane and mediate the flow of ions and metabolites between the cytosol and the mitochondrial intermembrane space.[2] VDAC and GRP78 are major players in the physical association between the endoplasmic reticulum (ER) and mitochondria which enables highly efficient transmission of Ca^{2+} from the ER to mitochondria under both physiological and pathological conditions.[29] GRP78 expression is induced under cerebral ischemia and serves as a protector in both in vivo and in vitro models of strokes.[30,31] GRP78 is specifically induced in cells under the unfolded protein response (UPR) and exported from the ER to mitochondria where it associates with VDAC in the outer mitochondrial membrane.[32] GRP78 is also expressed at the plasma membrane where it associates with VDAC.[33]

THE VOLTAGE-DEPENDENT ANION CHANNEL AND BETA-AMYLOID PEPTIDES

VDAC on the plasma membrane is also an amyloid Aβ peptide receptor. The Aβ mono- and/or oligomers dock to VDAC1 on the surface of neuronal cells via GxxxG motifs inducing channel opening that is followed by activation of the extrinsic apoptotic pathway.[34] In the neuron surface, VDAC1 is abundant in caveolae lipid rafts and commonly associated with two membrane-related isoforms of estrogen receptor alpha (mERalpha) (80 and 67 kDa) which participate in estrogen-induced neuroprotection against Aβ peptide injury[35] via a mechanism which involves VDAC1 phosphorylation that maintains the channel in an unactivated (closed) state.[36] In AD, the exposure of neurons to Aβ peptides enhances the dephosphorylation of VDAC1 which contributes to the progression of the disease.[37] As stated earlier, a second mechanism, also involving VDAC, inhibits the dephosphorylation of VDAC induced by Aβ peptides via Pg K5, which promotes phosphorylation at VDAC Ser12 and Thr107 through upregulation of GSK-3β.[27] The amyloid Aβ peptides also mediate several events leading to apoptosis including detachment of hexokinase, induction of VDAC oligomerization, and cytochrome c release.[38] The toxicity of the Aβ peptides involves both mitochondrial and plasma membrane VDAC1.[38]

DISORDERS ASSOCIATED TO DYSFUNCTIONS OF THE SYSTEM INVOLVING GRP78 AND VDAC

As a result of the UPR, GRP78 expression is induced and exported from the ER to mitochondria where it associates with VDAC in the outer mitochondrial membrane,[32] or in the cell surface.[39] Furthermore, ER stress can cause subpopulations of GRP78 to redistribute from the ER lumen to the cytosol and the ER membrane.[40] GRP78 prevents apoptosis that arises from disturbance of intracellular Ca^{2+}, suggesting the critical role of GRP78 in regulating cellular Ca^{2+} homeostasis.[32]

Overexpression of GRP78 in cells inhibits the production of Aβ peptides[41] via a mechanism where GRP78 binding to the amyloid precursor protein (APP) inhibits the translocation of APP from the ER to the Golgi apparatus, preventing the secretase-dependent proteolytic processing of APP.[42] It is well known that GRP78 impairs the aggregation of proteins in the ER.[43]

Intracellular aggregation of Aβ plays an important role in the pathogenesis of AD.[44] Analyzes in individual postmortem brain regions shows significantly increased VDAC1 expression in frontal cortical tissues from all AD patients relative to control subjects.[45] VDAC1 increases with age in cerebral cortex tissue and is correlated with high levels of Aβ peptides in AD patients and rat hippocampal neurons obtained after injection of Aβ peptides.[45] The Aβ injection damaged the mitochondrial structure of hippocampal neurons, decreased ATPase activity and mitochondrial membrane potential, and increased intracellular Ca^{2+}, increasing the dysfunctions leading to neurodegenerative effects.[45] A significant fraction of the VDAC1 overexpressed in the hippocampus of amyloidogenic AD transgenic mice was phosphorylated at an epitope susceptible to phosphorylation by GSK-3β, whose activity was also increased.[46] Furthermore, the levels of HX-I, which interacts with VDAC and affects its function, were decreased in mitochondria from AD models, indicating that reduced HX-I levels favor a VDAC1 involvement in disease progression of AD.[46]

Plasma membrane VDAC (pl-VDAC) is also involved in AD pathogenesis.[47] Estrogen receptor alpha (mERα) and pl-VDAC are present in the caveolae tissues from the human cortex and hippocampus, in a complex with scaffolding caveolin-1 which likely provides mERα stability at the plasma membrane.[47] In AD brains, pl-VDAC was accumulated in the caveolae and in dystrophic neurites of senile plaques and mERα was expressed in astrocytes surrounding the plaques, suggesting that pl-VDAC is involved in membrane dysfunction similar to that observed in AD neuropathology.[47]

VDAC AS A RECEPTOR OF THE PLASMINOGEN ACTIVATOR SYSTEM IN THE BRAIN

Pg together with its physiological activator, tissue-type Pg activator (t-PA), are primarily involved in fibrinolysis,[48] however, they also play critical nonfibrinolytic functions in the CNS.[49] VDAC is a receptor for Pg and t-PA.[50] Also, VDAC on the cell surface is associated with GRP78, and both are bridged together when Pg and t-PA bind to the cell surface.[33] Pg binds to a VDAC domain between amino acids Gly^{244}–Leu^{257} via its K5 domain, and t-PA binds to a VDAC domain between Gly^{20}–Gly^{24}. Pg binding to VDAC induces closure of the channel,[51] an effect enhanced by the expression of t-PA in the brain.[26] K5 binds to both VDAC1 and GRP78 on the cell surface, but only VDAC1 initiates a Ca^{2+} signaling cascade.[52] Conversely, GRP78 mediates a Ca^{2+} signaling cascade when it binds microplasminogen, the C-terminal region of Pg containing the protease domain. The binding of K5 involves an influx of Ca^{2+} from the extracellular environment toward the cytosol, whereas microplasminogen induces a release of Ca^{2+} from intracellular stores.[52] K5 is rapidly released from the surface after reduction by the NADH-dependent oxidoreductase activity of VDAC, whereas t-PA can be displaced by amyloid β peptides from the surface. These amyloid β peptides have the potential to aggregate and bind to VDAC leading to apoptosis. However, Pm prevents their aggregation via cleavage of β-(1–42) between Arg5 and His6, a site in amyloid β known to promote cell adhesion,[53] limiting their aggregation while maintaining the capacity of the amyloid β peptide to displace t-PA from VDAC. We constructed a model to show these mechanisms (Fig. 5.1). Because VDAC on the cell surface is also associated with GRP78,[33] we expanded the above model to include this association (Fig. 5.2) showing the four intracellular domains of GRP78.[54] The colocalization of these proteins permits the proximity of the GRP78 extracellular domains to VDAC, where the GRP78 N-terminal domain Leu^{98}–Leu^{115} is the t-PA binding site,[33] and the

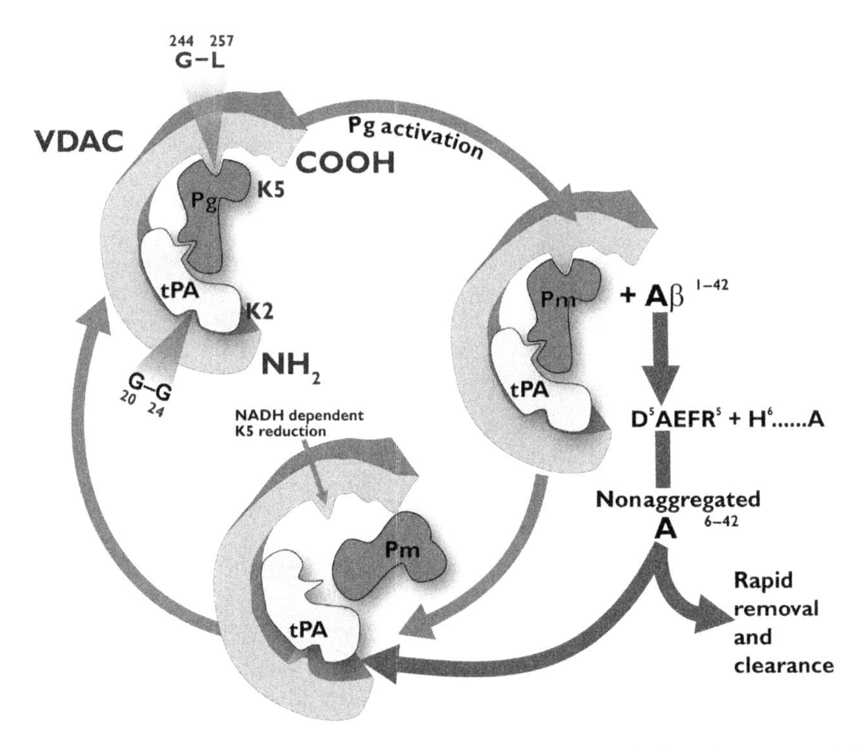

Figure 5.1 Interactions of VDAC with t-PA and Pg on the cell surface. VDAC binds t-PA and Pg via sites including amino acids Gly^{20}–Gly^{24} and Gly^{244}–Leu^{257}, respectively. This interaction promotes activation of Pg by t-PA. The Pm generated proteolyzes the Aβ 1–42 peptide which competes with t-PA for binding to VDAC. The Pm-truncated Aβ 1–42 peptide is unable to aggregate and is rapidly cleared from the receptor surface. Additionally, Pm can be released from the receptor through reduction of its K5 via the NADH-dependent oxidoreductase activity of VDAC.

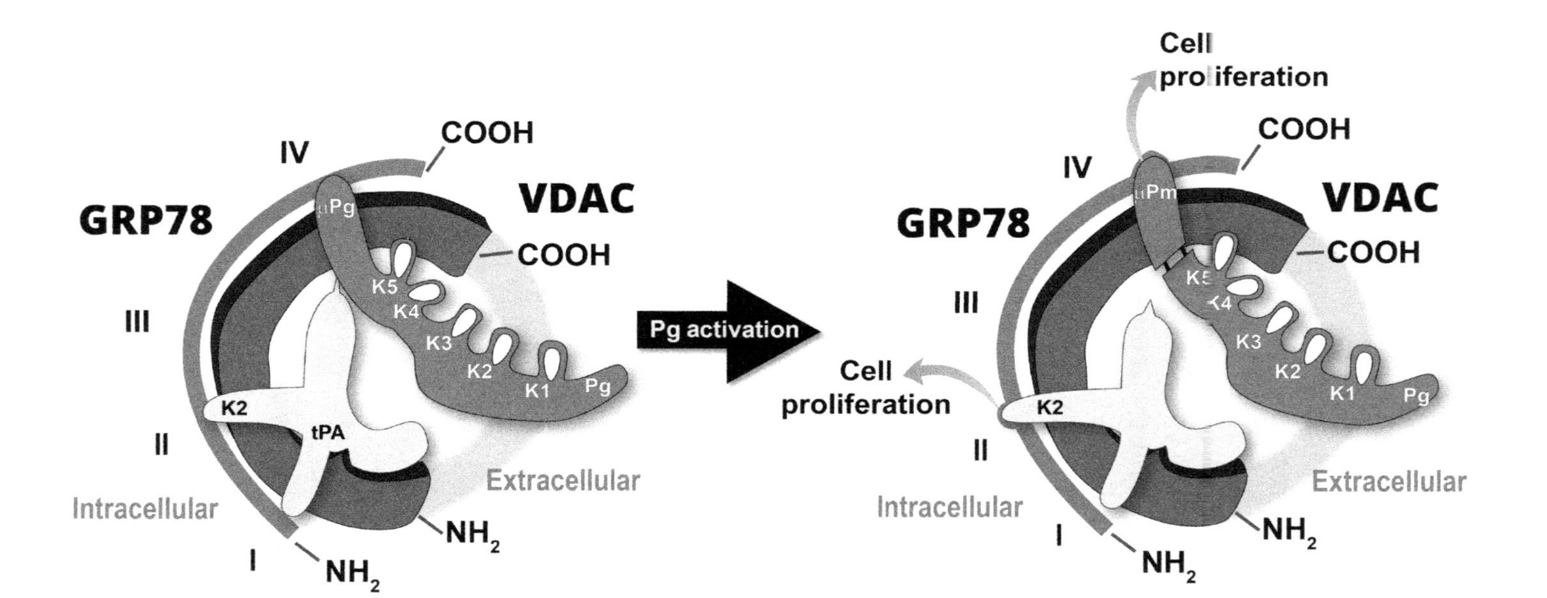

Figure 5.2 Interaction of colocalized GRP78 and VDAC with t-PA and Pg on the cell surface. Both t-PA and Pg bind to VDAC or GRP78. t-PA binds via its Kringle 2 to the GRP78 $Leu^{98}-Leu^{115}$ NH_2-terminal domain and Pg binds via its microplasminogen domain to the $Lys^{633}-Asp^{647}$ at the GRP78 COOH-terminal domain. Both t-PA and microplasminogen, once converted into microplasmin, promote cell proliferation.

GRP78 C-terminal region $Lys^{633}-Asp^{647}$ is the microplasminogen binding site.[52] t-PA binds to $GRP78-Lys^{113}$ in a lysine-dependent manner,[33] inducing a conformational change in t-PA, that facilitates its lysine-independent binding to VDAC. In addition to stimulating cell proliferation via GRP78, binding of t-PA to VDAC may induce further enhancement of Pg activation on the cell surface. Similarly, Pg binds to VDAC via its K5 which facilitates its conversion to Pm.[55] K5 may inhibit cell growth, however, its role is short-lived because it may be inactivated after reduction by the NADH-dependent activity of VDAC on the surface, although binding of the microplasminogen/microplasmin domain to the GRP78 C-terminal domain may induce a further increase of cell proliferation.[33] Because t-PA is very abundant in the brain[56] and because this organ is usually devoid of fibrinogen, GRP78 and VDAC mimic fibrin to enhance the activation of locally synthesized Pg, which can then act on nonfibrin substrates[49] such as VDAC itself. Therefore, this fibrin-independent mechanism may shift the conformational equilibrium of the protease domain of t-PA to the fully active form, and may also be necessary to promote brain cell proliferation and survival.

REFERENCES

1. Colombini M. A candidate for the permeability pathway of the outer mitochondrial membrane. *Nature* 1979;**279**:388–94.
2. Sampson MJ, Lovell RS, Craigen WJ. The murine voltage-dependent anion channel gene family. Conserved structure and function. *J Biol Chem* 1997;**272**:18966–73.
3. De Pinto V, Messina A, Acardi R, Aiello R, Guarino F, Tommasello MF, et al. New functions of an old protein: the eukaryotic porin or voltage-dependent anion selective channel (VDAC). *Ital J Biochem* 2003;**52**:17–24.
4. McEnery MW, Dawson TM, Verma A, Gurley D, Colombini M, Snyder SH. Mitochondrial voltage-dependent anion channel. *J Biol Chem* 1993;**268**:23289–96.
5. Shoshan-Barmatz V, De Pinto V, Zweckstetetter M, Raviv Z, Keinan N, Ariel N. VDAC, a multi-functional mitochondrial protein regulating cell life and death. *Mol Aspects Med* 2010;**31**:227–85.
6. Elinder F, Akanda N, Tofighi R, Shimizu S, Tsujinoto Y, Orrenius S, et al. Opening of plasma membrane voltage-dependent anion channels (VDAC) precedes caspase activation in neuronal apoptosis induced by toxic stimuli. *Cell Death Differ* 2005;**12**:1134–40.
7. Abu-Hama S, Zaid H, Israelson A, Nahon E, Shoshan-Barmatz V. Hexokinase-1 protection against apoptotic cell death is mediated via interaction with the voltage-dependent anion channel-1: mapping the site of binding. *J Biol Chem* 2008;**283**:13482.
8. Abu-Hamad S, Arbel N, Calo D, Arzoine L, Israelson A, Keinan N, et al. The VDAC1 N-terminus is essential both for apoptosis and the protective effect of anti-apoptotic proteins. *J Cell Sci* 2009;**122**:1906–16.

9. Pastorino JG, Hoek JB. Regulation of hexokinase binding to VDAC. *J Bioenerg Biomembr* 2008;**40**:171–82.

10. Shimizu S, Narita M, Tsujimoto Y. Bcl-2 family proteins regulate the release of apoptogenic cytochrome c by the mitochondrial channel VDAC. *Nature* 1999;**399**:483–7.

11. Elstrom RL, Bauer DE, Buzzai M, Karnauskas R, Harris MH, Plas DR, et al. Akt stimulates aerobic glycolysis in cancer cells. *Cancer Res* 2001;**64**:3892–9.

12. Miyamoto S, Murphy AN, Brown JH. Akt mediates mitochondrial protection in cardiomyocytes through phosphorylation of mitochondrial hexokinase-II. *Cell Death Differ* 2008;**15**:521–9.

13. Pastorino JG, Hoek JB, Shulga N. Activation of glycogen synthase kinase-3beta disrupts the binding of hexokinase II to mitochondria by phosphorylating voltage-dependent anion channel and potentiates chemotherapy-induced cytotoxicity. *Cancer Res* 2005;**65**:10545–54.

14. Pastorino JG, Shulga N, Hoek JB. Mitochondrial binding of hexokinase II inhibits Bax-induced cytochrome c release and apoptosis. *J Biol Chem* 2002;**277**:7610–18.

15. Linseman DA, Butts BD, Precht TA, Phelps RA, Le SS, Laessig TA, et al. Glycogen synthase kinase3beta phosphorylates Bax and promotes its mitochondrial localization during neuronal apoptosis. *J Neurosci* 2004;**24**:9993–10002.

16. Yamamoto T, Yamada A, Watanabe M, Yoshimura Y, Yamazaki N, Yoshimura Y, et al. VDAC1, having a shorter N-terminus than VDAC2 but showing the same migration in an SDS-polyacrylamide gel, is the predominant form expressed in mitochondria of various tissues. *J Proteome Res* 2006;**5**:3336–44.

17. Ghosh T, Pasndey N, Maitra A, Brahmachari SK, Pillai B. A role for voltage-dependent anion channel VDAC1 in polyglutamine-mediated neuronal cell death. *PLoS One* 2007;**2**: e1170.

18. Poleti MD, Tesch AC, Crepaldi CR, Souza GH, Eberlin MN, de Cerqueira CM. Relationship between expression of voltage-dependent anion channel (VDAC) isoforms and type of hexokinase binding sites on brain mitochondria. *J Mol Neurosci* 2010;**41**:48–54.

19. Imada S, Yamamoto M, Tanaka K, Seiwa C, Watanabe K, Kamei Y, et al. Hypothermia-induced increase of oligodendrocyte precursor cells: Possible involvement of plasmalemmal voltage-dependent anion channel 1. *J Neurosci Res* 2010;**88**:3457–66.

20. Melchor JP, Strickland S. Tissue plasminogen activator in central nervous system physiology and pathology. *Thromb Haemost* 2005;**93**:655–60.

21. Samson AL, Medcalf RL. tissue-type plasminogen activator: a multifaceted modulator of neurotransmission and synaptic plasticity. *Neuron* 2006;**50**:673–8.

22. Gur-Wanon D, Mizrachi T, Maaravi-Pinto FY, Lourbopoulos A, Grigoriadis N, Higazi AA, et al. The plasminogen activator system: involvement in central nervous system inflammation and a potential site for therapeutic intervention. *J Neuroinflammation* 2013;**10**:124–35.

23. Sheehan JJ, Tsirka SE. Fibrin-modifying serine proteases thrombin, tPA and plasmin in ischemic stroke: a review. *Glia* 2005;**50**:340–50.

24. Kawashita E, Kanno Y, Asayama H, Okada K, Ueshima S, Matsuo O, et al. Involvement of α_2-antiplasmin in dendritic growth of hippocampal neurons. *J Neurochem* 2013;**126**:59–69.

25. Soeda S, Koyanagi S, Kuramoto Y, Kimura M, Oda M, Kozako T, et al. Anti-apoptotic roles of plasminogen activator inhibitor-1 as a neurotrophic factor in the central nervous system. *Thromb Haemost* 2008;**100**:1014–20.

26. Li J, Yu L, Gu X, Ma Y, Pasqualini R, Arap W, et al. Tissue plasminogen activator regulates Purkinje neuron development and survival. *Proc Natl Acad Sci USA* 2013;**110**: E2410–19.

27. Li L, Yao YC, Gu XQ, Che D, Ma CQ, Dai ZY, et al. Plasminogen kringle 5 induces endothelial cell apoptosis by triggering a voltage-dependent anion channel 1 (VDAC1) positive feedback loop. *J Biol Chem* 2014;**289**:32628–38.

28. Soeda S, Oda M, Ochiai S, Shimeno H. Deficient release of plasminogen activator inhibitor-1 from astrocytes triggers apoptosis in neuronal cells. *Brain Res Mol Brain Res* 2001;**91**:96–103.

29. Ouyang YB, Giffard RG. ER-mitochondria crosstalk during cerebral ischemia: molecular chaperones and ER-Mitochondrial calcium transfer. *Int J Cell Biol* 2012;**2012**:1–8.

30. Ouyang YB, Lu Y, Yue S, Xu LJ, Xiong XX, White RE, et al. MiR181 regulates GRP78 and influences outcome from cerebral ischemia in vitro and in vivo. *Neurobiol Dis* 2012;**45**:555–63.

31. Rajdev S, Hara K, Kokubo Y, Mestril R, Dillmann W, Weinstein PR, et al. Mice overexpressing rat heat shock protein 70 are protected against cerebral infarction. *Ann Neurol* 2000;**47**:782–91.

32. Sun FC, Wei S, Li CW, Chang YS, Chao CC, Lai YK. Localization of GRP78 to mitochondria under the unfolded protein response. *Biochem J* 2006;**396**:31–9.

33. Gonzalez-Gronow M, Gomez CF, de Ridder GG, Ray R, Pizzo SV. Binding of tissue-type plasminogen activator to the glucose-regulated protein 78 (GRP78) modulates plasminogen activation and promotes human neuroblastoma cell proliferation in vitro. *J Biol Chem* 2014;**5**:25166–76.

34. Thinnes FP. Plasmalemmal VDAC-1 corroborated as amyloid Aβ-receptor. *Front Aging Neurosci* 2015;**7**:188–93.

35. Marin R, Ramírez CM, González M, González-Muñoz E, Zorzano A, Camps M, et al. Voltage-dependent anion channel (VDAC) participates in amyloid beta-induced toxicity and interacts with plasma membrane estrogen receptor alpha in septal and hippocampal neurons. *Mol Memb Biol* 2007;**24**:148–60.

36. Herrera JL, Díaz M, Hernández-Fernaud JR, Salido E, Alonso M, Fernández C, et al. Voltage-dependent anion channel as a resident protein of lipid rafts: post-transductional regulation by estrogens and involvement in neuronal preservation against Alzheimer's disease. *J Neurochem* 2011;**116**:820–7.

37. Fernández-Echevarría C, Díaz M, Ferrer I, Canerina-Amaro A, Marin R. Aβ promotes VDAC1 channel dephosphorylation in neuronal lipid rafts. Relevance to the mechanisms of neurotoxicity in Alzheimer's disease. *Neuroscience* 2014;**278**:354–66.

38. Smilansky AQ, Dangoor L, Nakdimon I, Ben-Hail D, Shoshan-Barmatz V. The voltage dependent anion channel 1 mediates amyloid β toxicity and represents a potential target for Alzheimer disease therapy. *J Biol Chem* 2015;**290**:30670–83.

39. Delpino A, Castelli M. The 70 kDa glucose-regulated protein (GRP78/BIP) is expressed on the cell membrane, is released into cell culture medium and is also present in human peripheral circulation. *Biosci Rep* 2002;**22**:407–20.

40. Rao RV, Peel A, Logvinova A, del Rìo G, Hermel E, Yokota T, et al. Coupling endoplasmic reticulum stress to the cell death program: role of the ER chaperone GRP78. *FEBS Lett* 2002;**514**:122–8.

41. Yang Y, Turner RFS, Gaut JR. The chaperone BIP/GRP78 bind to amyloid precursor protein and decreases Aβ40 and Aβ42 secretion. *J Biol Chem* 1998;**273**:25552–5.

42. Hoshino T, Nakaya T, Araki W, Suzuki K, Suzuki T, Mizushima T. Endoplasmic reticulum chaperones inhibit the production of amyloid-β peptides. *Biochem J* 2007;**402**:581–9.

43. Lee AS. The glucose-regulated proteins: stress induction and clinical applications. *Trends Biochem Sci* 2001;**26**:504–10.

44. Hartmann T. Intracellular biology of Alzheimer's disease amyloid β peptide. *Eur Arch Psychiatry Clin Neurosci* 1999;**249**:291–8.

45. Reddy PH. Is the mitochondrial outer membrane protein VDAC1 therapeutic target for Alzheimer's disease? *Biochim Biophys Acta* 2013;**1832**:67–75.

46. Cuadrado-Tejedor M, Vilariño M, Cabodevilla F, Del Río J, Pérez-Mediavilla A. Enhanced expression of the voltage-dependent anion channel 1 (VDAC1) in Alzheimer's disease transgenic mice: an insight into the pathogenic effects of amyloid-β. *J Alzheimers Dis* 2011;**23**:195–206.

47. Ramírez CM, González M, Díaz M, Alonso R, Ferrer I, Santpere G, et al. VDAC and ERalpha interaction in caveolae from human cortex is altered in Alzheimer's disease. *Mol Cell Neurosci* 2009;**42**:172–83.

48. Hoylaerts M, Rijken DC, Lijnen HR, Collen D. Kinetics of the activation of plasminogen by human tissue plasminogen activator. Role of fibrin. *J Biol Chem* 1982;**257**:2912–19.

49. Tsirka SE, Bugge TH, Dewgen JL, Strickland S. Neuronal death in the central nervous system demonstrates a non-fibrin substrate for plasmin. *Proc Natl Acad Sci USA* 1997;**94**:9779–81.

50. Gonzalez-Gronow M, Ray R, Wang F, Pizzo SV. The voltage-dependent anion channel (IVDAC) binds tissue-type plasminogen activator and promotes activation of plasminogen on the cell surface. *J Biol Chem* 2013;**288**:498–509.

51. Banerjee J, Ghosh S. Interaction of mitochondrial voltage-dependent anion channel from rat brain with plasminogen protein leads to partial closure of the channel. *Biochim Biophys Acta* 2004;**1663**:6–8.

52. Gonzalez-Gronow M, Kaczowka SJ, Payne S, Wang F, Gawdi G, Pizzo SV. Plasminogen structural domains exhibit different functions when associated with cell surface GRP78 or the voltage-dependent anion channel. *J Biol Chem* 2007;**282**:32811–20.

53. Ghiso J, Rostagno A, Gardella JE, Liem L, Gorevic PD, Frangione B. A 109-amino acid C-terminal fragment of Alzheimer's disease amyloid precursor protein contains a sequence, RHDS, that promotes cell adhesion. *Biochem J* 1992;**288**:1053–9.

54. Gonzalez-Gronow M, Selim MA, Papalas J, Pizzo SV. GRP78: a multifunctional receptor on the cell surface. *Antioxid Redox Signal* 2009;**11**:2299–306.

55. Gonzalez-Gronow M, Kalfa T, Johnson CE, Gawdi G, Pizzo SV. The voltage-dependent anion channel is a receptor for plasminogen kringle 5 on human endothelial cells. *J Biol Chem* 2003;**278**:27312–18.

56. Davies BJ, Pickard BS, Steel M, Morris RG, Lathe R. Serine proteases in rodent hippocampus. *J Biol Chem* 1998;**273**:23004–11.

Cell Surface GRP78: A Targetable Marker of Cancer Stem-Like Cells

Robin E. Bachelder
Duke University Medical Center, Durham, NC, United States

DRUG-RESISTANT, GROWTH-ARRESTED TUMOR CELL SUBPOPULATIONS DRIVE TUMOR RECURRENCE

Current cancer therapies target actively proliferating tumor cells. While these treatment strategies effectively reduce tumor burden, residual chemoresistant tumor cells frequently persist and are responsible for tumor recurrence and patient mortality. Accordingly, development of more effective cancer therapies will be reliant on the recognition that tumors are heterogeneous, consisting of proliferating, therapy-sensitive

Cell Surface GRP78, a New Paradigm in Signal Transduction Biology. DOI: https://doi.org/10.1016/B978-0-12-812351-5.00006-4
© 2018 Elsevier Inc. All rights reserved.

tumor cells and growth-arrested, drug-resistant tumor cell subpopulations.[1–3] Treatment with drugs targeting proliferating tumor cells reduces overall tumor burden, but enriches for the nonproliferating, drug-resistant tumor cell subpopulations, which are frequently more invasive and metastatic than the bulk tumor[4] (Fig. 6.1). These invasive populations can accordingly drive tumor recurrence, which is ultimately responsible for patient mortality. This knowledge underscores the need to identify targetable signaling pathways in drug-resistant, invasive tumor cell subpopulations. Our laboratory has identified novel resistance pathways by enriching for these tumor cell subpopulations using a short-term chemotherapy treatment model.[1,4,5]

EPITHELIAL–MESENCHYMAL TRANSITION AND CANCER STEMNESS

Numerous studies show that drug-resistant, invasive tumor cell subpopulations have undergone an epithelial–mesenchymal transition (EMT), and exhibit properties of stem cells.[6–10] In order to identify molecular pathways that may be targeted to eliminate residual disease, it is important to understand the biology of EMT and cancer stem cells.

The EMT is thought to reflect the process of tumor progression, namely tumor cell acquisition of invasive and metastatic properties, as

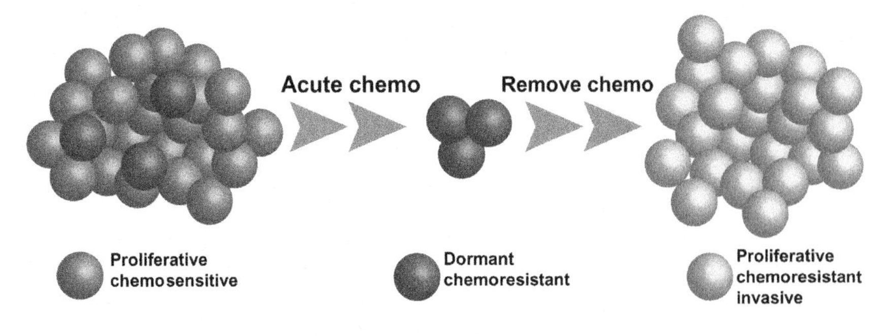

Figure 6.1 Tumor heterogeneity and implications for therapy resistance.
Tumors are heterogeneous, consisting of proliferative chemosensitive tumor cells (blue (gray in print version)), and dormant, chemoresistant cells (red (dark gray in print version)). Notably, chemoresistant tumor cells can be represented at a low frequency in the bulk tumor. Our laboratory has demonstrated that short-term chemotherapy treatment enriches for the chemoresistant subpopulation, allowing for identification of novel signaling axes that drive resistance.[1,5] We have also published the important finding that after removing chemotherapy, chemoresidual tumors resume proliferation and acquire a more invasive phenotype driven by their acquisition of novel proinvasive signaling pathways.[4] Our collective work suggests that tumors can acquire a more invasive/metastatic phenotype during lapses in treatment, which likely drives local and distant tumor recurrence.

well as therapy resistance. This transition involves the loss of expression of epithelial adhesion proteins and acquired expression of mesenchymal proteins that promote an invasive phenotype. These alterations in gene expression are orchestrated by a set of transcriptional repressors and activators, including Snail-1, Slug, and Twist. Overexpression of such transcriptional repressors in normal breast epithelial cells drives EMT, cancer stemness, and drug resistance, underscoring a central role for these proteins in tumor progression.[10]

Cancer stem/progenitor cells are defined by their ability to: (1) self renew through asymmetric division, and (2) undergo multilineage differentiation. Self-renewing activity is measured by growing cancer cells in nonadherent tissue culture conditions in the presence of defined growth factors (EGF, bFGF). Under these conditions, cancer stem cells form spheres. Cells from these spheres, when dissociated, are able to generate daughter spheres, thus exhibiting self-renewing activity.[11] Notably, a fraction of these cells when provided with a physiologic environment, can form glands with defined basal and luminal layers, demonstrating their ability to undergo multilineage differentiation.[11,12] Cancer stem cells, compared to bulk tumor cells, also exhibit enhanced tumor initiating activity, as measured by their ability when injected at limiting cell numbers to efficiently form tumors in animal models.[13]

Relevant to the topic of this chapter, cancer stem cells exhibit therapy resistance.[14] Frequently, cancer stem-like cells express ATP binding cassette transporters, which effectively pump therapies out of cells, resulting in therapy resistance. Likewise, cancer stem-like cells express bcl-2 family members such as survivin that maintain cell survival in the face of toxic insults. Finally, the altered metabolic profile of cancer stem-like cells can contribute to their therapy resistance.

GLUCOSE REGULATED PROTEIN OF 78 KDA (GRP78), STEMNESS, AND THERAPY RESISTANCE

In addition to being an endoplasmic reticulum chaperone protein, GRP78 is expressed on the surface of a variety of tumor types where it orchestrates numerous signaling pathways.[15] GRP78 knockout studies indicate an essential role of this protein in maintaining survival of embryonic[16] and adult[17] mammary stem cells. Cell surface GRP78

(CS-GRP78) protects mammary stem cells from apoptosis by binding to Cripto, a GRP78 ligand.[17]

Research from our laboratory and others indicates that CS-GRP78 is a marker of cancer stem-like cells. The first group to identify an association between CS-GRP78 and cancer stemness performed cell sorting to show that CS-GRP78(+) head and neck cancer cells exhibited significantly elevated sphere forming potential relative to CS-GRP78(−) head and neck cancer cells.[18] Moreover, our laboratory showed that ascites from ovarian cancer patients enriches for a CS-GRP78(+) ovarian cancer cell subpopulation which exhibits cancer stem-like properties, including the ability to form spheres.[19] By cell sorting we have also shown that CS-GRP78(+) prostate cancer cells exhibit enhanced prostasphere growth compared to CS-GRP78(−) prostate cancer cells (unpublished data).

GRP78 ligands play a central role in cancer stem cell activities. Cripto is a cell surface glycoprotein that serves as an obligatory coreceptor for the transforming growth factor-B. Importantly, this Cripto activity is absolutely dependent on its binding to GRP78.[20] Cripto maintains stem cell pluripotency and GRP78 antibodies that block Cripto binding (i.e., antibodies directed against the GRP78 N-terminus) eliminate stem cells.

CS-GRP78 -expressing tumor cells are frequently therapy resistant.[21] CS-GRP78 sorted head and neck tumors exhibit increased resistance to both radiation and chemotherapy. Likewise, we have shown that ascites from ovarian cancer patients enriches for multidrug resistant ovarian cancer stem-like cells that express CS-GRP78.[19,22]

TARGETING CELL SURFACE GRP78 ON DRUG-RESISTANT CANCER STEM-LIKE CELLS

CS-GRP78 is expressed in cancer cells, but not in normal cells, making it an ideal target for cancer therapy.[23,24] Significant effort has been invested in targeting CS-GRP78 to eliminate therapy-resistant cancer stem-like cells.[19,20,25] The laboratory of Dr. Salvatore Pizzo has produced and characterized a unique set of monoclonal antibodies specific for different domains of CS-GRP78. Studies from this laboratory indicate that ligands that bind to the amino-terminus (N-terminus) of CS-GRP78 activate Akt.[26] Conversely, antibodies that bind to the carboxyl-terminus (C-terminus) of GRP78 inhibit Akt in prostate tumor cells.[27] Furthermore, we showed that a C-terminal GRP78 monoclonal antibody eliminates ascites-enriched ovarian cancer stem-like cells by inhibiting Akt activity.[19] This finding is important because it shows that this C-terminal antibody eliminates ovarian cancer stem-like cells, and when combined with chemotherapy has the potential to prevent tumor recurrence. Similarly, our unpublished studies show that this C-terminal GRP78 antibody binds to prostate cancer stem-like cells, and suppresses prostasphere growth.

Akt phosphorylates Glycogen Synthase Kinase-3 (GSK-3), resulting in suppression of GSK-3 activity. GSK-3 is a key regulator of Snail-l, a transcriptional repressor that drives stem-like cell behavior.[10,28] This collective knowledge suggests that an Akt/GSK-3/Snail-1 signaling axis may be a central determinant of cancer stem-like cell behavior. Our published studies indicate that a carboxy-terminal GRP78 monoclonal antibody eliminates ovarian cancer stem cell growth by inhibiting an Akt/GSK-3/Snail-1 signaling axis.[19] Importantly, we have demonstrated similar efficacy of this antibody in eliminating prostate cancer stem-like cells by inhibiting this signaling axis (unpublished data).

PTEN is a tumor suppressor that is deleted in multiple tumor types.[29,30] Thus, PTEN deletion is oncogenic. PTEN is also a central repressor of Akt activity. Considering that cancer stem cells are dependent on Akt for their survival, PTEN suppresses cancer stem cell activities. A seminal study showed that conditional PTEN deletion in the mouse prostate results in development of metastatic prostate cancer. Furthermore, double knockout studies indicate that oncogenicity of PTEN deletion, and subsequent activation of Akt, is lost upon coincident GRP78 deletion in prostate epithelium.[29] These studies provide compelling evidence that the tumor suppressive activity of PTEN is completely dependent on its suppression of a GRP78/Akt signaling axis in cancer stem-like cells. These findings underscore the importance of testing efficacy of targeting CS-GRP78 to eliminate prostate cancer stem cells in PTEN-deleted tumors.

CANCER STEM CELLS AND THEIR TRANSIT AMPLIFYING PROGENY- DIFFERENCES IN PROLIFERATIVE POTENTIAL

Cancer stem cells transition through several stages depending on the availability of relevant microenvironmental cues. Pluripotent stem cells are quiescent, whereas transit amplifying cells acquire the ability to proliferate. These transit amplifying cells represent a transitional state between pluripotency and differentiation. In the context of the cancer stem cell field, quiescent cancer stem cells exhibit elevated tumor initiating activity when injected into orthotopic models. In contrast, transit amplifying cancer stem cells can proliferate, but do not exhibit tumor initiating activity because they have lost the ability to self-renew.

Different markers are expressed in pluripotent cells, transit amplifying cells, and differentiated cells. A number of transcription factors are expressed in pluripotent stem cells (e.g., Sox2, c-myc, Oct4, Lin28), and are thought to maintain a state of dedifferentiation. Notably, transition of quiescent stem cells to proliferating, transit amplifying cells is dependent on the expression of Sonic Hedgehog, a ligand of the hedgehog signaling pathway, which is implicated in tumor progression.[31–35] CS-GRP78 expression has been reported in both normal and cancer stem cells, as well as in transit amplifying cells.[36] Intriguingly, Dr. Pizzo's laboratory has shown that C-terminal GRP78 antibodies inhibit cancer cell proliferation.[37–39] However, this laboratory has also shown that CS-GRP78-sorted cancer cells are quiescent.[19] These collective findings suggest that CS-GRP78 is expressed in both dedifferentiated and differentiated cancer cells, reflecting its potential involvement in multiple stages of tumor progression.

THE ONE-TWO PUNCH: COMBINATION THERAPIES TO ELIMINATE THERAPY-SENSITIVE AND THERAPY-RESISTANT TUMOR CELL SUBPOPULATIONS

Our studies and others indicate that tumors are heterogeneous, consisting of therapy-sensitive and therapy-resistant tumor cell subpopulations—the latter being responsible for tumor recurrence (see Fig. 6.1). According to this model, monotherapy will never be an effective treatment for cancer patients. Instead, we need to develop effective combination therapies that target these heterogeneous tumor cell subpopulations (see Fig. 6.2). A central focus of our laboratory is to test the efficacy of combining standard of care therapies (targeting proliferating tumor cells) with a C-terminal GRP78 monoclonal antibody (targeting cancer stem-like cells) to more effectively target heterogeneous cancer populations in women's (breast, ovarian) and men's (prostate) cancers, thus preventing drug resistance and tumor recurrence after standard of care treatment.

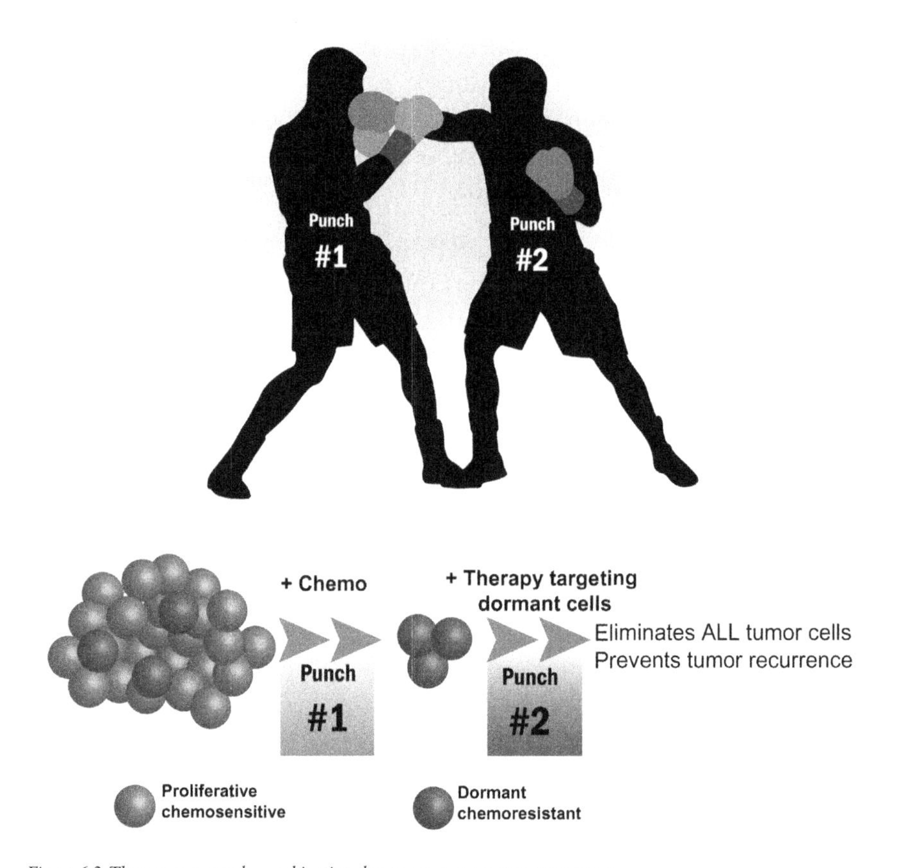

Figure 6.2 The one-two punch: combination therapy to prevent tumor recurrence.
Tumors consist of both therapy-sensitive and therapy-resistant tumor cell subpopulations. Treatments that target only proliferative cells debulk the tumor, but residual drug-resistant tumor cells ultimately result in tumor recurrence and patient mortality. Accordingly, combination therapies that target these heterogeneous tumor cell subpopulations ("punch #1" and "punch #2") are needed to prevent tumor recurrence and prolong patient survival. Chemoresistant tumor cells frequently exhibit properties/signaling of cancer stem-like cells. We have identified a tumor specific marker (cell surface GRP78) on drug-resistant prostate and ovarian cancer cells that drives therapy resistance and cancer stem-like behaviors.[19,26] Moreover, we have shown that monoclonal antibodies binding to the carboxyl terminal domain of cell surface GRP78 induce apoptosis of these drug-resistant tumor cell subpopulations. These findings underscore the importance of preclinical studies investigating efficacy of combination therapy (chemotherapy + C-terminal GRP78 antibody) in eliminating heterogeneous cancer cell subpopulations, thus preventing tumor recurrence.

ARE GRP78+ STEM-LIKE CELLS INDUCED BY THERAPY OR ENRICHED FOR BY THERAPY, AND WHY DO WE CARE?

While long-term exposure of cancer cells to chemotherapy promotes acquired resistance by driving resistance mutations, numerous studies indicate that heterogeneous tumors are also composed of cells with intrinsic therapy resistance.[1–5] Exposure of these heterogeneous tumors to short-term therapy enriches for tumor cell subpopulations with intrinsic therapy resistance. Markers of resistance identified in these models are also expressed in primary patient tumors.[4] Importantly, exposure of these heterogeneous tumors to therapy enriches for therapy-resistant tumor cells, but subsequent removal of therapy drives reversion to a heterogeneous tumor.[1,3]

Our studies indicate that CS-GRP78(+) tumor cells are represented at a low frequency in heterogeneous ovarian cancers, and are enriched by ascites both in vitro and in ovarian cancer patients.[19] Notably, we also showed in vitro that ascites-enriched GRP78(+) tumor cells revert back to the original heterogeneous tumor phenotype after ascites removal. To provide evidence that these CS-GRP78(+) stem-like cancer cells were preexisting in the original tumor, we demonstrated by cell sorting that CS-GRP78(+) tumor cells sorted from the bulk ovarian cancer cell population exhibited enhanced sphere forming ability and increased tumor initiating activity compared to the CS-GRP78(−) sorted cells.[19] Similar sorting experiments in head and neck cancers demonstrated that a preexisting subpopulation of these head and neck cancers expresses GRP78, and is therapy resistant.[18]

Why is it important to distinguish between intrinsic and acquired therapy resistance? Development of the most effective treatment strategies for cancer patients will rely on determining whether therapy-resistant tumor cells are a preexisting subpopulation of cells in the bulk tumor. If CS-GRP78(+) tumor cells are detected in heterogeneous primary tumors, we suggest that GRP78-targeted therapies should be administered to patients prior to standard of care therapy in order to prevent therapy-dependent enrichment of these aggressive tumor cells. In contrast, if CS-GRP78(+) tumor cells are produced only after therapy administration, then GRP78 inhibitors should be administered after completion of these therapies.

CONCLUSIONS

Cancers are heterogeneous, consisting of therapy-resistant and therapy-sensitive tumor cells with different invasive and proliferative potential. Accordingly, while monotherapy can debulk patient tumors, residual therapy-resistant cancer cells are responsible for tumor recurrence and patient mortality. This collective knowledge indicates that combination therapies are needed to reduce patient mortality. Therapy-resistant cancer cells exhibit properties of cancer stem-like cells, and support signaling pathways associated with the EMT. One logical therapeutic target for resistant cells is CS-GRP78. This protein is expressed on the surface of therapy-resistant tumor cell subpopulations in both women's and men's cancers, and it is implicated in cancer stem-like behaviors. Antibodies that bind to the carboxyl terminal epitopes of CS-GRP78 eliminate therapy-resistant cancer stem-like cells by inhibiting an Akt signaling axis. These findings underscore the need to study efficacy of combination therapy, including both standard of care therapies targeting proliferative cells and a GRP78 C-terminal antibody that targets therapy-resistant stem-like cells.

REFERENCES

1. Li S, Kennedy M, Payne S, Kennedy K, Seewaldt VL, Pizzo SV, et al. Model of tumor dormancy/recurrence after short-term chemotherapy. *PLoS One* 2014;**9**:e98021.
2. Moore N, Houghton J, Lyle S. Slow-cycling therapy-resistant cancer cells. *Stem Cells Dev* 2012;**21**:1822–30.
3. Sharma SV, Lee DY, Li B, Quinlan MP, Takahashi F, Maheswaran S, et al. A chromatin-mediated reversible drug-tolerant state in cancer cell subpopulations. *Cell* 2010;**141**:69–80.
4. Nelson ER, Li S, Kennedy M, Payne S, Kilibarda K, Groth J, et al. Chemotherapy enriches for an invasive triple-negative breast tumor cell subpopulation expressing a precursor form of N-cadherin on the cell surface. *Oncotarget* 2016;**7**:84030–42.
5. Li S, Payne S, Wang F, Claus P, Su Z, Groth J, et al. Nuclear basic fibroblast growth factor regulates triple-negative breast cancer chemo-resistance. *Breast Cancer Res* 2015;**17**:91.
6. Achuthan S, Santhoshkumar TR, Prabhakar J, Nair SA, Pillai MR. Drug-induced senescence generates chemoresistant stemlike cells with low reactive oxygen species. *J Biol Chem* 2011;**286**:37813–29.
7. Bhola NE, Balko JM, Dugger TC, Kuba MG, Sanchez V, Sanders M, et al. TGF-beta inhibition enhances chemotherapy action against triple-negative breast cancer. *J Clin Invest* 2013;**123**:1348–58.
8. Calcagno AM, Salcido CD, Gillet JP, Wu CP, Fostel JM, Mumau MD, et al. Prolonged drug selection of breast cancer cells and enrichment of cancer stem cell characteristics. *J Natl Cancer Inst* 2010;**102**:1637–52.
9. Samanta D, Gilkes DM, Chaturvedi P, Xiang L, Semenza GL. Hypoxia-inducible factors are required for chemotherapy resistance of breast cancer stem cells. *Proc Natl Acad Sci USA* 2014;**111**:E5429–5438.
10. Mani SA, Guo W, Liao MJ, Eaton EN, Ayyanan A, Zhou AY, et al. The epithelial-mesenchymal transition generates cells with properties of stem cells. *Cell* 2008;**133**:704–15.
11. Dontu G, Abdallah WM, Foley JM, Jackson KW, Clarke MF, Kawamura MJ, et al. In vitro propagation and transcriptional profiling of human mammary stem/progenitor cells. *Genes Dev* 2003;**17**:1253–70.
12. Xin L, Lukacs RU, Lawson DA, Cheng D, Witte ON. Self-renewal and multilineage differentiation in vitro from murine prostate stem cells. *Stem Cells* 2007;**25**:2760–9.

13. Al-Hajj M, Wicha MS, Benito-Hernandez A, Morrison SJ, Clarke MF. Prospective identification of tumorigenic breast cancer cells. *Proc Natl Acad Sci USA* 2003;**100**:3983–8.

14. Economopoulou P, Kaklamani VG, Siziopikou K. The role of cancer stem cells in breast cancer initiation and progression: potential cancer stem cell-directed therapies. *Oncologist* 2012;**17**:1394–401.

15. Lee AS. The Par-4-GRP78 TRAIL, more twists and turns. *Cancer Biol Ther* 2009;**8**:2103–5.

16. Luo S, Mao C, Lee B, Lee AS. GRP78/BiP is required for cell proliferation and protecting the inner cell mass from apoptosis during early mouse embryonic development. *Mol Cell Biol* 2006;**26**:5688–97.

17. Spike BT, Kelber JA, Booker E, Kalathur M, Rodewald R, Lipianskaya J, et al. CRIPTO/GRP78 signaling maintains fetal and adult mammary stem cells ex vivo. *Stem Cell Rep* 2014;**2**:427–39.

18. Chiu CC, Lee LY, Li YC, Chen YJ, Lu YC, Li YL, et al. GRP78 as a therapeutic target for refractory head-neck cancer with CD24(−)CD44(+) stemness phenotype. *Cancer Gene Ther* 2013;**20**:606–15.

19. Mo L, Bachelder RE, Kennedy M, Chen PH, Chi JT, Berchuck A, et al. Syngeneic murine ovarian cancer model reveals that ascites enriches for ovarian cancer stem-like cells expressing membrane GRP78. *Mol Cancer Ther* 2015;**14**:747–56.

20. Kelber JA, Panopoulos AD, Shani G, Booker EC, Belmonte JC, Vale WW, et al. Blockade of Cripto binding to cell surface GRP78 inhibits oncogenic Cripto signaling via MAPK/PI3K and Smad2/3 pathways. *Oncogene* 2009;**28**:2324–36.

21. Zhang Y, Tseng CC, Tsai YL, Fu X, Schiff R, Lee AS. Cancer cells resistant to therapy promote cell surface relocalization of GRP78 which complexes with PI3K and enhances PI(3,4,5)P3 production. *PLoS One* 2013;**8**:e80071.

22. Mo L, Pospichalova V, Huang Z, Murphy SK, Payne S, Wang F, et al. Ascites increases expression/function of multidrug resistance proteins in ovarian cancer cells. *PLoS One* 2015;**10**:e0131579.

23. Arap MA, Lahdenranta J, Mintz PJ, Hajitou A, Sarkis AS, Arap W, et al. Cell surface expression of the stress response chaperone GRP78 enables tumor targeting by circulating ligands. *Cancer Cell* 2004;**6**:275–84.

24. Jakobsen CG, Rasmussen N, Laenkholm AV, Ditzel HJ. Phage display derived human monoclonal antibodies isolated by binding to the surface of live primary breast cancer cells recognize GRP78. *Cancer Res* 2007;**67**:9507–17.

25. Liu R, Li X, Gao W, Zhou Y, Wey S, Mitra SK, et al. Monoclonal antibody against cell surface GRP78 as a novel agent in suppressing PI3K/AKT signaling, tumor growth, and metastasis. *Clin Cancer Res* 2013;**19**:6802–11.

26. Misra UK, Deedwania R, Pizzo SV. Activation and cross-talk between Akt, NF-kappaB, and unfolded protein response signaling in 1-LN prostate cancer cells consequent to ligation of cell surface-associated GRP78. *J Biol Chem* 2006;**281**:13694–707.

27. Sato M, Yao VJ, Arap W, Pasqualini R. GRP78 signaling hub a receptor for targeted tumor therapy. *Adv Genet* 2010;**69**:97–114.

28. Bachelder RE, Yoon SO, Franci C, de Herreros AG, Mercurio AM. Glycogen synthase kinase-3 is an endogenous inhibitor of Snail transcription: implications for the epithelial-mesenchymal transition. *J Cell Biol* 2005;**168**:29–33.

29. Fu Y, Wey S, Wang M, Ye R, Liao CP, Roy-Burman P, et al. Pten null prostate tumorigenesis and AKT activation are blocked by targeted knockout of ER chaperone GRP78/BiP in prostate epithelium. *Proc Natl Acad Sci USA* 2008;**105**:19444–9.

30. Wey S, Luo B, Tseng CC, Ni M, Zhou H, Fu Y, et al. Inducible knockout of GRP78/BiP in the hematopoietic system suppresses Pten-null leukemogenesis and AKT oncogenic signaling. *Blood* 2012;**119**:817–25.

31. Hsu YC, Li L, Fuchs E. Transit-amplifying cells orchestrate stem cell activity and tissue regeneration. *Cell* 2014;**157**:935–49.

32. Fan L, Pepicelli CV, Dibble CC, Catbagan W, Zarycki JL, Laciak R, et al. Hedgehog signaling promotes prostate xenograft tumor growth. *Endocrinology* 2004;**145**:3961–70.

33. Sheng T, Li C, Zhang X, Chi S, He N, Chen K, et al. Activation of the hedgehog pathway in advanced prostate cancer. *Mol Cancer* 2004;**3**:29.

34. Karhadkar SS, Bova GS, Abdallah N, Dhara S, Gardner D, Maitra A, et al. Hedgehog signalling in prostate regeneration, neoplasia and metastasis. *Nature* 2004;**431**:707–12.

35. Carpenter RL, Lo HW. Hedgehog pathway and GLI1 isoforms in human cancer. *Discov Med* 2012;**13**:105–13.

36. Heijmans J, van Lidth de Jeude JF, Koo BK, Rosekrans SL, Wielenga MC, van de Wetering M, et al. ER stress causes rapid loss of intestinal epithelial stemness through activation of the unfolded protein response. *Cell Rep* 2013;**3**:1128–39.

37. de Ridder GG, Ray R, Pizzo SV. A murine monoclonal antibody directed against the carboxyl-terminal domain of GRP78 suppresses melanoma growth in mice. *Melanoma Res* 2012;**22**:225–35.

38. de Ridder G, Ray R, Misra UK, Pizzo SV. Modulation of the unfolded protein response by GRP78 in prostate cancer. *Methods Enzymol* 2011;**489**:245–57.

39. Misra UK, Pizzo SV. Ligation of cell surface GRP78 with antibody directed against the COOH-terminal domain of GRP78 suppresses Ras/MAPK and PI 3-kinase/AKT signaling while promoting caspase activation in human prostate cancer cells. *Cancer Biol Ther* 2010;**9**:142–52.

CHAPTER 7

Escherichia coli Subtilase Cleaves Cell Surface GRP78 Preventing COOH-Terminal Domain Signaling

Rupa Ray
Duke University Medical Center, Durham, NC, United States

GRP78

Glucose Regulated Protein 78,000 (GRP78) is a well-characterized molecular chaperone of the HSP70 family. It is primarily located in the luminal space of the endoplasmic reticulum (ER) of all cell types where it binds to nascent polypeptides and is responsible for sequestering these proteins until they can be properly folded and/or oligomerized. In the ER, GRP78 is also the master regulator of the unfolded protein response (UPR), a cellular stress response related to ER stress. When unfolded or misfolded proteins accumulate in the lumen of the ER, the UPR is triggered, and has the following three primary functions: (1) To restore normal cellular function by halting protein translation,

Cell Surface GRP78, a New Paradigm in Signal Transduction Biology. DOI: https://doi.org/10.1016/B978-0-12-812351-5.00007-6
© 2018 Elsevier Inc. All rights reserved.

(2) degrade misfolded proteins, and (3) increase the production of proteins involved in protein translation via the activation of relevant signaling pathways. During the UPR, GRP78 dissociates from the UPR mediators PKR-like ER kinase (PERK), inositol-requiring enzyme 1(IRE1), and activating transcription factor 6 (ATF6), activating downstream pathways to decrease protein synthesis and increase chaperone transcription to allow GRP78 to participate in protein folding. The UPR ultimately acts to promote either survival or apoptotic pathways in response to ER stress. Normal cellular function is, therefore, reestablished by reducing intermediate protein aggregates, increasing protein folding, regulation of Ca^{2+}, and the repression of translation.[1] In addition to its role as a molecular chaperone in the ER, GRP78 is found in the cytoplasm, nucleus, mitochondria, and in secreted and plasma membrane-associated forms.[2–6] Despite its carboxyl-terminal ER-retention signal (KDEL), GRP78 is translocated to the cell surface under certain physiologic conditions (e.g., hypoxia),[7,8] and GRP78 is expressed on the surface of a variety of cells, ranging from cancer cells[9] and ECs[10] to activated macrophages.[11]

GRP78 at the cell surface is detected in multiple cancer cells in vivo and is usually not present on nonmalignant cells.[9,12] Furthermore, the expression of GRP78 on the cell surface is associated with tumor progression, metastasis, and a poor prognosis.[13] Research from Dr. Pizzo's laboratory described the acceleration of murine melanoma growth by autoantibodies similar to those found in human cancer patients.[14] This selective expression of GRP78 on multiple types of tumors makes it a particularly enticing anticancer therapeutic target.

Cell surface GRP78 (CS-GRP78) activates different pathways depending on whether the ligand recognizes the NH_2-terminal domain (NTD) or the COOH-terminal domain (CTD). Activated α_2M^* binds to GRP78 NTD and stimulates survival and proliferation in a number of cancer cell types.[15,16] The binding of α_2M^* to 1-LN prostate cancer cells promotes their proliferation in a MAPK and Akt-dependent manner.[17] Autoantibodies that recognize an epitope in the NTD of GRP78 often occur in prostate cancer,[18] ovarian cancer,[19] and melanoma[14] as well as being correlated with poor prognosis. In a manner antagonistic to the NTD signaling, exogenous CTD-reactive antibodies upregulate p53 and promote apoptosis in prostate cancer cells.[20] These observations demonstrate that GRP78 has diverse roles in addition to its function as a chaperone in the ER.

SUBTILASE

The subtilase cytotoxin (SubAB) represents the fourth and newest family of AB5 toxins. It is produced by certain virulent strains of Shiga toxigenic *Escherichia coli* (STEC) and was first isolated from O113:21 strain of STEC that caused an outbreak of hemolytic uremic syndrome (HUS) in South Australia in 1998.[21] Since this initial identification, SubAB has been detected in a variety of STEC serotypes.[22] SubAB comprises a 35 kDa catalytic A subunit (SubA) and 5 13 kDa B subunits (SubB). The A subunit contains the catalytic triad Asp, His, and Ser, and mutation of any of these three resides results in a catalytically inactive enzyme. The A subunit is a serine protease, and this activity is necessary for its cytotoxic effects. The B subunit mediates binding to glycan receptors on the cell surface and is necessary for triggering internalization and subsequent trafficking of the holotoxin to the ER.[21] Moreover, this occurs via a clathrin-dependent process and not via lipid rafts.[23] Interestingly, SubB binds preferentially to a nonhuman glycan, α2-3 linked *N*-glycolylneuraminic acid (Neu5Gc).[24] This glycan is not synthesized in humans, however, it is obtained from the consumption of food derived from mammals that endogenously produce it.

The SubA serine proteinase displays extreme substrate specificity, and its only identified substrate is the ER chaperone GRP78. This specificity is because the A subunit has an unusually deep active site cleft, unlike most other subtilase enzymes. SubA cleaves GRP78 between the amino acid residues Leu^{416}–Leu^{417}. This cleavage site is in the hinge region between the ATPase and COOH-terminal protein binding domains, and cleavage at this site leads to the production of a 28 kDa COOH-terminal fragment.[25] Because of its critical role in maintaining protein homeostasis, cleavage of GRP78 induces cell death. Furthermore, the exquisite substrate specificity of SubA for GRP78 suggests that it is a valuable tool for probing the role(s) of GRP78.

SUBTILASE AS A TOOL FOR UNDERSTANDING GRP78 FUNCTIONS

Because GRP78 is present in many cellular compartments (ER, cytoplasm, nucleus, mitochondria, secreted, cell surface),[26] it is difficult to study the function of this protein within any one compartment.

Current methods to reduce GRP78 expression (RNAi knockdown, knockout mice) reduce global GRP78 expression regardless of the cellular compartment of interest. Moreover, GRP78 knockout mice are embryonic lethal,[27] and although conditional/temporal knockouts have been developed,[28] they do not address individual cellular compartments. RNAi knockdown of GRP78 also reduces the entire pool of GRP78 in a cell, and 100% knockdown is never attained, further complicating studies to determine the function of GRP78 in specific contexts. However, SubAB treatment of cells results in rapid degradation of cellular GRP78 pools, mimicking a knockout phenotype.[25] In particular, the Pizzo laboratory showed that SubA specifically cleaves CS-GRP78, resulting in abrogation of the effects of CTD-reactive antibodies to GRP78, without compromising the function of the NTD-reactive antibodies.[29] This specificity of SubA for GRP78, as well as its inability to be internalized, provides an elegant tool for studying CS-GRP78 without interference from intracellular processes dependent on GRP78, and SubA cleavage of CS-GRP78 makes it possible to study the receptor functions of GRP78 without nonspecific effects caused by knockdown/cleavage of ER GRP78. Additionally, because ER GRP78 is the master regulator of the UPR, SubAB provides a specific tool with which to further elucidate the molecular mechanism underlying UPR induction and its resulting consequences.

UNFOLDED PROTEIN RESPONSE

As discussed earlier, because GRP78 is essential for the UPR, and because it is the only substrate of SubAB, the most well-described effects of SubAB are on the UPR and subsequent downstream events. In the first paper describing the interaction of GRP78 and SubAB, the Paton laboratory showed that the protease activity of SubA was dependent on Ser272 because a mutation to Alanine did not cause GRP78 cleavage. Moreover, they showed that mice injected with SubAB showed activation of the UPR in the liver, but that mice treated with Shiga toxin, a related AB5 toxin, did not undergo UPR.[25] This was a key finding in that no other cytotoxin is known to directly target chaperone proteins or ER components. This study laid the foundation for subsequent studies that further elucidated the mechanism by which SubAB cleavage of GRP78 triggers the UPR, which eventually culminates in apoptosis.

Consistent with its role in UPR induction, SubAB was shown to inhibit protein synthesis by Morinaga et al.,[30] and a follow-up study by the same group determined that this was caused by cleavage of ER GRP78, thereby inducing the UPR.[31] UPR induction triggered the phosphorylation of PERK and eIF2α, components of one of the three UPR signaling pathways downstream of GRP78, resulting in transient inhibition of protein translation. Moreover, SubAB-treated Vero cells showed cell cycle arrest at G1 and inhibition of DNA synthesis.[32] Additional studies in Vero cells confirmed the activation of PERK and phosphorylation of eIF2α. Furthermore, activation of the other two UPR signaling pathways, IRE1 and ATF6, was also demonstrated, consistent with the hypothesis that GRP78 cleavage by SubAB causes it to dissociate from UPR mediators, thereby initiating the UPR.[33]

INFLAMMATION

Nuclear factor κB (NF-κB), a protein complex controlling DNA transcription, cytokine production, and cell survival interacts with GRP78, although the role GRP78 plays in NF-κB activation remains unclear.[17] NF-κB is also important for the pathology of HUS and renal injury, and in this context, SubAB cleavage of GRP78 causes Akt phosphorylation and subsequent NF-κB activation via the ATF6 branch of the UPR during the early phase.[34] Conversely, pretreatment of mice with sublethal doses of SubAB also triggered the UPR and protected against lipopolysaccharide inflammatory injury. However, it caused suppression of NF-κB and NF-κB gene expression.[35] Further experiments showed that the mechanism for this suppressive effect of SubAB on NF-κB β was due to activation of the ATF6 branch of the UPR, which caused induction of C/EBPβ (CCAAT/enhancer-binding protein), a transcription factor with important roles in cell proliferation and differentiation, as well as suppression of Akt.[36] These studies suggest that GRP78 plays a role in both NF-κB and Akt-dependent diseases, including cancer, cardiovascular disease, diabetes, inflammatory diseases, Parkinson's disease, and Alzheimer's disease.

Playing key roles in inflammation are the proteins intercellular cell adhesion molecule-1 (ICAM-1) and vascular cell adhesion molecule-1 (VCAM-1). Both are expressed on endothelial cell (EC) surfaces and are important for the movement of polymorphonuclear leukocytes from the blood to sites of inflammation, and interleukin-8 (IL-8) and monocyte chemotactic protein-1 (MCP-1), which target leukocytes or monocytes to sites of inflammation, respectively, and their expression is under transcriptional control of NF-κB. To determine the role of ER stress—and specifically GRP78—in EC inflammation, human pulmonary vein endothelial cells were treated with SubAB to deplete ER GRP78. Results showed decreased NF-κB binding to DNA, resulting in the reduced expression of ICAM-1, VCAM-1, IL-8, and MCP-1. This also caused reorganization of the actin cytoskeleton and restoration of endothelial permeability.[37] Similarly, SubAB was used to determine the mechanism of action of geranylgeranylacetone, an antiulcer agent.[38] A putative mechanism of antiinflammatory action for indomethacin in podocytes was elucidated involving GRP78, NF-κB, and tumor necrosis factor receptor-associated protein 2 (TRAF2).[39] Taken together, these studies show that GRP78 is important in inflammation responses via its interaction with NF-κB.

APOPTOSIS

Although activation of the UPR should restore cellular homeostasis, in some cases failure to sufficiently overcome this stress can lead to apoptosis. This UPR-induced apoptosis is mediated by the same molecules and pathways that are responsible for maintenance of ER. This balance is maintained by complex regulation of these three UPR signaling pathways, PERK, IRE1, and ATF6, and UPR-induced apoptosis occurs as a result of cytochrome c release, caspase activation and DNA fragmentation.[40] Similarly, SubAB cleavage of GRP78 ultimately causes cell death, both in vitro and in vivo.

Prolonged ER stress causes PERK phosphorylation of eIF2α and induction of ATF4 expression, resulting in CHOP up-regulation, a transcription factor that regulates proapoptotic gene expression. This increased expression of CHOP leads to decreased Bcl-2 (apoptosis regulator Bcl-2) levels and translocation of Bax (Bcl-2 antagonist of cell death) to the mitochondria to induce apoptosis. Consistent with this, SubAB primarily activates apoptotic pathways involved by Bax/Bak (Bcl-2 associated X protein), leading to mitochondrial membrane damage.[41] This confirms the role of these proteins in GRP78-mediated cell death.

Expanding on this observation, Yahiro et al., showed that on HeLa cells, neural/glial antigen 2 (NG2), $\alpha 2\beta 1$ integrin (ITG), L1 cell adhesion molecule (L1CAM), and hepatocyte growth factor receptor (Met) are receptors of SubAB and trigger Bax/Bak mediated apoptosis.[42] They further showed that death associated protein 1 (DAP1) is required for Bax/Bak-dependent apoptosis after SubAB cleavage of GRP78 and that this is dependent on the PERK-eIF2α pathway. Additionally, activation of PERK suppresses autophagy, a catabolic process that degrades cytosolic components, also via DAP1. This may occur through the interaction of DAP1 and mTOR (Ser/Thr kinase mammalian target of rapamycin).[43] Notably, mTOR and GRP78 interact in ER-stress induced autophagy,[44] and DAP1 could be a link in this molecular mechanism. Additionally, CS-GRP78 ligation by its natural ligand, α2M, results in activation of the mTORC1 and mTORC2 complexes, of which mTOR is a critical component, and promotes protein synthesis in prostate cancer cells, although autophagy was not evaluated.[45,46] However, these observations suggest that ligation of CS-GRP78 may lead to autophagy via the resulting activation of mTORC1 and mTORC2 and their possible interaction with DAP1.

CANCER

As discussed elsewhere is this book, GRP78, both ER and cell surface, plays critical roles in tumor proliferation, progression, and metastasis. The high specificity of SubAB for GRP78 suggests that it may be used as therapeutic strategy to cause ER-stress induced apoptosis. Indeed, treatment of human melanoma cells (CHL-1 and WM266-4) with SubAB not only caused cell death when used as the sole agent, but also increased the sensitivity of these cells to fenretinide or bortezomib.[46] Although SubAB does not show specificity for cancer cells, sublethal doses of SubAB combined with vemurafenib, melphalan, and temozolomide did increase the sensitivity of human melanoma cells line to these drugs, regardless of BRAF or PTEN status (Ray, unpublished data).

However, to circumvent the uptake of SubAB by both cancer and noncancer cells, fusion proteins of SubA (the catalytic subunit) and ligands for receptors found mainly on cancer cells have been developed. The first such fusion protein was with EGF (epidermal growth factor), the ligand for EGFR (epidermal growth factor receptor), which is frequently overexpressed in tumor cells. They showed that this engineered construct is highly toxic to EGRF expressing growing and confluent cells and that this is caused by cleavage of ER GRP78. In human breast and prostate tumor xenograft mouse models, the construct also inhibited tumor growth. Finally, the EGF-SubA sensitized cancer cells to ER-stress inducing drugs.[47] This construct was also used in glioblastoma cell lines. Consistent with the melanoma results, EGF-SubA exhibited tumor-specific proteolytic activity and cytotoxicity in glioblastoma cell lines and potentiated the antitumor activity of both temozolomide and ionizing radiation. Moreover, the fusion construct was tolerated and led to tumor growth delay in a glioma xenograft mouse model.[48]

In a pair of studies to determine the effects of EGF-SubA combined with photodynamic therapy (PDT), a treatment comprising systemic/local administration of a photosensitizer followed by tumor illumination with visible light efficiently killed cancer cells by triggering an atypical nonapoptotic death accompanied by high levels of cytoplasmic vacuolation in cell culture lines.[49] In vivo, EGF-SubA also improved the efficacy of PDT; however, the dose was lower so as not to impair the immune system.[50]

In a comparable approach, Zhang et al., constructed a fusion protein in which the B subunit was replaced with the GRP78 binding peptide (GBP). Because GRP78 is highly expressed on tumor cells, but not nonmalignant cells, this construct specifically targets cancer cells. Moreover, the GBP was previously shown to target and bind tumor-specific CS-GRP78. Similar to EGF-SubA, GBP-SubA both targeted and killed cells expressing GRP78.[51] These studies all indicate that cleavage of ER GRP78 using a tumor-specific marker for targeting and SubA for cleavage is a novel strategy for the development of anti-tumor drugs. Additionally, these constructs may sensitize cancer cells to traditional chemotherapies, radiation, or PDT, which may lead to a reduction in dosage of the treatments, resulting in fewer side effects.

VIRUSES

CS-GRP78 acts as a receptor or coreceptor for a number of viruses, such as the flavivirus Dengue (DENV),[52] and ER GRP78 is involved in the pathogenesis of many viral diseases, including hepatitis B[53] and respiratory syncytial virus.[54] SubAB cleavage of ER GRP78 was shown to halt cytomegalovirus virion production in the cytoplasm of infected cells, indicating a significant role for GRP78 in virion formation and cytoplasmic egress.[55] In addition to its role as a coreceptor for DENV, GRP78 is also important for DENV viral antigen production. SubAB cleavage of ER GRP78 in DENV-infected cells led to the loss of intracellular DENV particles, and a significant decrease in intracellular DENV antigen, although DENV RNA levels were unchanged, indicating normal DENV RNA replication.[56] Therefore, GRP78 appears to be necessary for both DENV infection and viral antigen production and accumulation. Japanese encephalitis virus (JEV), another flavivirus, also requires GRP78 for both viral entry and replication. Here, SubAB was used to demonstrate that GRP78 is a novel host factor involved in several steps of the JEV life cycle and may be a therapeutic target.[57] As GRP78 becomes implicated in the pathogenesis of more viral diseases, due to its important role in protein folding and likely viral entry and replication, SubAB will be useful for further elucidating its function.

CELL SURFACE GRP78

As discussed previously, GRP78 acts as a growth factor-like receptor at the cell surface and is involved in both proliferation and apoptosis, depending on where it is ligated. There has been much controversy regarding the topology of CS-GRP78 and the A subunit of Subtilase has been a powerful tool for revealing the structure and function of CS-GRP78.

Ray et al.,[29] showed that CS-GRP78 is specifically cleaved by the catalytic A subunit of the subtilase cytotoxin SubAB, and that CS-GRP78 cleavage does not affect intracellular GRP78 expression levels. Cleavage of the COOH-terminal 28 kDa of GRP78 abrogates apoptosis caused by ligation of CTD antibodies to GRP78, and cleavage of GRP78 by SubA results in a decrease in Ca^{2+} signaling induced by CTD antibody binding to GRP78. Cleavage of CS-GRP78 by SubA has broad implications, not only for the study of GRP78, but also regarding its role on the cell surface in cancers and infectious diseases.

Notably, SubA only cleaves CS-GRP78, likely because if the B subunit is present, it will direct rapid binding to receptors on the surface to trigger uptake of the holotoxin, preventing the SubA interaction with surface GRP78. Moreover, any holotoxin that remains bound to its glycan receptor cannot interact with GRP78.

Previous studies using transmembrane prediction programs suggested that GRP78 has four potential transmembrane domains at the cell surface: I (a.a. 1–17), II (a.a. 29–45), III (a.a. 222–242), and IV (a.a. 414–431).[58] However, based on these studies, at least residues 416–417 are exposed on the cell surface, because this is the sole SubA cleavage site, and cell-surface cleavage resulting in a 29 kDa COOH-terminal fragment detectable in cell culture supernatants is observed. Although predicted transmembrane domains I through III, and part of IV, may be actual transmembrane domains, the first three residues of predicted transmembrane domain IV are exposed to the extracellular space.

Cleavage of the CTD of CS-GRP78 also results in the abrogation of signaling initiated by ligation of GRP78 by CTD antibodies. However, signaling due to ligation of anti-NTD GRP78 antibodies is unaffected. Anti-CTD antibodies to GRP78 induce apoptosis via a p53 dependent mechanism,[20] while also decreasing the expression of the antiapoptotic protein Bcl-2 and increasing the expression of the proapoptotic proteins BAD, BAX, and BAK.[59] The levels of the cleaved caspases 3, 7, 8, and 9 are also increased when CS-GRP78 is ligated with CTD antibody.[59] Conversely, NTD agonists promote cell survival and proliferation through activation of PI-3k and Akt.[59]

Previous studies of Ca^{2+} release induced by the CTD antibody C20 and the NTD antibody N20 in the highly metastatic prostate cancer cell line 1-LN showed that treatment with C20 resulted in a sustained Ca^{2+} release from intracellular stores, whereas treatment with N20 resulted in a brief spike of Ca^{2+} release.[60] Similar wave forms were observed with HepG2 cells, such that C20 treatment resulted in a prolonged signal compared to N20 treatment. However, the C20-induced Ca^{2+}release is abrogated by cleavage of the COOH-terminus of GRP78 by SubA, whereas Ca^{2+} release by the NTD-reactive N20 antibody is unaffected in SubA-treated cells. This suggests that the COOH-terminus of GRP78 does not interact with the NH_2-terminus of GRP78, and that it is not necessary for signaling through the NH_2-terminus, indicating that the two termini act as separate signaling compartments.

This proposed selective ligand determination of differential GRP78 signaling is not completely without precedent. Evidence from the 7-transmembrane domain and adrenergic receptor fields indicate that different adrenergic drugs can exhibit "biased agonism" when interacting with a single receptor type. The immediate branch-point in this system is selectivity for signaling via G-proteins or β-arrestins, with significantly different biological sequelae. This has been explored both in vitro and in vivo using the β_2-adrenergic receptors and angiotensin type 1 receptors as models.[61–63] Furthermore, regional bias in signaling has been demonstrated using the protease-activated receptors (PARs) in which distinct patterns of PAR coupling to effector molecules depend on differential regional interactions with proteinases. The result of this regional selectivity also leads to markedly different biological outcomes.[64]

SubA cleavage of CS-GRP78 confirms that both the NH2-terminus and COOH-terminus perform different functions. Proliferation induced by ligation of the NH2-terminus with NTD antibodies was not affected by CS-GRP78 cleavage, whereas apoptosis was abrogated by cleavage of the COOH-terminus. These data indicate that the CTD is indeed the region of GRP78 which is important for initiating apoptosis and confirms previous studies that indicate that the CTD of GRP78 is the site that transduces the apoptotic signal. The abrogation of signaling when cells are treated with SubA suggests that the cleavage of the CTD of GRP78 would be protective against apoptosis caused by potential circulating CTD-reactive antibodies.

As discussed earlier, GRP78 is a receptor for Dengue virus on HepG2 cells. Specifically, NTD antibodies inhibit Dengue virus entry, whereas CTD antibodies enhance virus entry.[52] This suggests that cleavage of CS-GRP78 by SubA, or otherwise interfering with CTD availability, may be a unique strategy for combating this disease. Moreover, the novel observation that SubA specifically cleaves CS-GRP78 suggests that it may be of great utility for both infectious disease treatment and prevention, as well as for specifically targeting cancer cells.

CONCLUSIONS

In summary, since its initial identification in 1998, SubAB has been used to probe the function of GRP78 in a number of different contexts. Its exquisite substrate specificity makes it particularly suited for teasing apart the various roles that GRP78 plays in a multitude of cellular contexts. Additionally, the finding that the A subunit specifically cleaves CS-GRP78 provides us with a unique tool to study the receptor function of GRP78 on a variety of cell types without disturbing the intracellular compartments where GRP78 performs functions necessary for cell survival. Moreover, SubAB, particularly because of its specificity for GRP78 may have applications for the clinical treatment of human diseases involving GRP78. Fig. 7.1 is a schematic diagram reflecting the contents of this chapter.

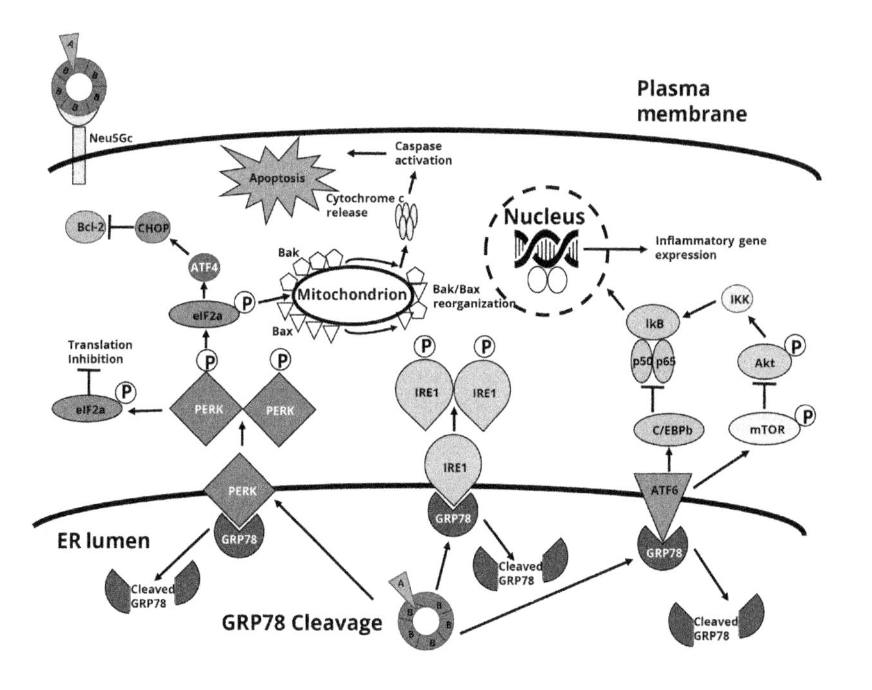

Figure 7.1 SubAB cleaves ER GRP78 and activates the UPR and induces inflammation and apoptosis. Subtilase binds to Neu5Gc on the cell surface. It undergoes clathrin-dependent retrograde transport via the Golgi network to the ER, where it cleaves GRP78. This causes GRP78 to disassociate from the UPR mediators PERK, IRE1, and ATF6. PERK phosphorylates the transcription factor eIF2α, leading to inhibition of protein translation. Prolonged ER stress also causes PERK phosphorylation of eIF2α and induction of ATF4 expression, resulting in CHOP up-regulation, a transcription factor that regulates proapoptotic gene expression, which leads to decreased Bcl-2 levels and translocation of Bax to the mitochondria to induce apoptosis. SubAB primarily activates apoptotic pathways triggered by Bax/Bak reorganization, leading to mitochondrial membrane damage, cytochrome c release, caspase activation, and apoptosis. For inflammation, SubAB cleavage of GRP78 causes Akt phosphorylation. This subsequently leads to NF-κB activation via the ATF6 branch of the UPR during the early phase, which causes induction of C/EBPβ (CCAAT/enhancer-binding protein), a transcription factor with important roles in cell proliferation, differentiation, as well as suppression of Akt. Finally, NF-κB translocates to the nucleus where it regulates the expression of proinflammatory genes, such as MCP-1, ICAM, VCAM, and IL-8.

REFERENCES

1. Quinones QJ, de Ridder GG, Pizzo SV. GRP78: a chaperone with diverse roles beyond the endoplasmic reticulum. *Histol Histopathol* 2008;**23**:1409–16.
2. Ni M, Zhou H, Wey S, Baumeister P, Lee AS. Regulation of PERK signaling and leukemic cell survival by a novel cytosolic isoform of the UPR regulator GRP78/BiP. *PLoS One* 2009;**4**:e6868.
3. Reddy RK, Mao C, Baumeister P, Austin RC, Kaufman RJ, Lee AS. Endoplasmic reticulum chaperone protein GRP78 protects cells from apoptosis induced by topoisomerase inhibitors: role of ATP binding site in suppression of caspase-7 activation. *J Biol Chem* 2003;**278**:20915–24.
4. Sun FC, Wei S, Li CW, Chang YS, Chao CC, Lai YK. Localization of GRP78 to mitochondria under the unfolded protein response. *Biochem J* 2006;**396**:31–9.
5. Kern J, Untergasser G, Zenzmaier C, Sarg B, Gastl G, Gunsilius E, et al. GRP-78 secreted by tumor cells blocks the antiangiogenic activity of bortezomib. *Blood* 2009;**114**:3960–7.
6. Berger CL, Dong Z, Hanlon D, Bisaccia E, Edelson RL. A lymphocyte cell surface heat shock protein homologous to the endoplasmic reticulum chaperone, immunoglobulin heavy chain binding protein BIP. *Int J Cancer* 1997;**71**:1077–85.
7. Tsai YL, Zhang Y, Tseng CC, Stanciauskas R, Pinaud F, Lee AS. Characterization and mechanism of stress-induced translocation of 78-kilodalton glucose-regulated protein (GRP78) to the cell surface. *J Biol Chem* 2015;**290**:8049–64.
8. Song MS, Park YK, Lee JH, Park K. Induction of glucose-regulated protein 78 by chronic hypoxia in human gastric tumor cells through a protein kinase C-epsilon/ERK/AP-1 signaling cascade. *Cancer Res* 2001;**61**:8322–30.
9. Arap MA, Lahdenranta J, Mintz PJ, Hajitou A, Sarkis AS, Arap W, et al. Cell surface expression of the stress response chaperone GRP78 enables tumor targeting by circulating ligands. *Cancer Cell* 2004;**6**:275–84.
10. Raiter A, Weiss C, Bechor Z, Ben-Dor I, Battler A, Kaplan B, et al. Activation of GRP78 on endothelial cell membranes by an ADAM15-derived peptide induces angiogenesis. *J Vasc Res* 2010;**47**:399–411.
11. Misra UK, Pizzo SV. Up-regulation of GRP78 and antiapoptotic signaling in murine peritoneal macrophages exposed to insulin. *J Leukoc Biol* 2005;**78**:187–94.
12. Kim Y, Lillo AM, Steiniger SC, Liu Y, Ballatore C, Anichini A, et al. Targeting heat shock proteins on cancer cells: selection, characterization, and cell-penetrating properties of a peptidic GRP78 ligand. *Biochemistry* 2006;**45**:9434–44.
13. Lee AS. GRP78 induction in cancer: therapeutic and prognostic implications. *Cancer Res* 2007;**67**:3496–9.
14. de Ridder GG, Gonzalez-Gronow M, Ray R, Pizzo SV. Autoantibodies against cell surface GRP78 promote tumor growth in a murine model of melanoma. *Melanoma Res* 2010;**21**:35–43.
15. Misra UK, Sharma T, Pizzo SV. Ligation of cell surface-associated glucose-regulated protein 78 by receptor-recognized forms of alpha 2-macroglobulin: activation of p21-activated protein kinase-2-dependent signaling in murine peritoneal macrophages. *J Immunol* 2005;**175**:2525–33.
16. Misra UK, Deedwania R, Pizzo SV. Binding of activated alpha2-macroglobulin to its cell surface receptor GRP78 in 1-LN prostate cancer cells regulates PAK-2-dependent activation of LIMK. *J Biol Chem* 2005;**280**:26278–86.

17. Misra UK, Deedwania R, Pizzo SV. Activation and cross-talk between Akt, NF-kappaB, and unfolded protein response signaling in 1-LN prostate cancer cells consequent to ligation of cell surface-associated GRP78. *J Biol Chem* 2006;**281**:13694–707.

18. Mintz PJ, Kim J, Do KA, Wang X, Zinner RG, Cristofanilli M, et al. Fingerprinting the circulating repertoire of antibodies from cancer patients. *Nat Biotechnol* 2003;**21**:57–63.

19. Cohen M, Petignat P. Purified autoantibodies against glucose-regulated protein 78 (GRP78) promote apoptosis and decrease invasiveness of ovarian cancer cells. *Cancer Lett* 2011;**309**:104–9.

20. Misra UK, Mowery Y, Kaczowka S, Pizzo SV. Ligation of cancer cell surface GRP78 with antibodies directed against its COOH-terminal domain up-regulates p53 activity and promotes apoptosis. *Mol Cancer Ther* 2009;**8**:1350–62.

21. Paton AW, Srimanote P, Talbot UM, Wang H, Paton JC. A new family of potent AB(5) cytotoxins produced by Shiga toxigenic *Escherichia coli*. *J Exp Med* 2004;**200**:35–46.

22. Nuesch-Inderbinen MT, Funk J, Cernela N, Tasara T, Klumpp J, Schmidt H, et al. Prevalence of subtilase cytotoxin-encoding subAB variants among Shiga toxin-producing *Escherichia coli* strains isolated from wild ruminants and sheep differs from that of cattle and pigs and is predominated by the new allelic variant subAB2-2. *Int J Med Microbiol* 2015;**305**:124–8.

23. Chong DC, Paton JC, Thorpe CM, Paton AW. Clathrin-dependent trafficking of subtilase cytotoxin, a novel AB5 toxin that targets the endoplasmic reticulum chaperone BiP. *Cell Microbiol* 2008;**10**:795–806.

24. Byres E, Paton AW, Paton JC, Lofling JC, Smith DF, Wilce MC, et al. Incorporation of a non-human glycan mediates human susceptibility to a bacterial toxin. *Nature* 2008;**456**:648–52.

25. Paton AW, Beddoe T, Thorpe CM, Whisstock JC, Wilce MC, Rossjohn J, et al. AB5 subtilase cytotoxin inactivates the endoplasmic reticulum chaperone BiP. *Nature* 2006;**443**:548–52.

26. Ni M, Zhang Y, Lee AS. Beyond the endoplasmic reticulum: atypical GRP78 in cell viability, signalling and therapeutic targeting. *Biochem J* 2011;**434**:181–8.

27. Luo S, Mao C, Lee B, Lee AS. GRP78/BiP is required for cell proliferation and protecting the inner cell mass from apoptosis during early mouse embryonic development. *Mol Cell Biol* 2006;**26**:5688–97.

28. Kim S, Wang M, Lee AS, Thompson RF. Impaired eyeblink conditioning in 78kDa-glucose regulated protein (GRP78)/immunoglobulin binding protein (BiP) conditional knockout mice. *Behav Neurosci* 2011;**125**:404–11.

29. Ray R, de Ridder GG, Eu JP, Paton AW, Paton JC, Pizzo SV. The *Escherichia coli* subtilase cytotoxin A subunit specifically cleaves cell-surface GRP78 protein and abolishes COOH-terminal-dependent signaling. *J Biol Chem* 2012;**287**:32755–69.

30. Yahiro K, Morinaga N, Satoh M, Matsuura G, Tomonaga T, Nomura F, et al. Identification and characterization of receptors for vacuolating activity of subtilase cytotoxin. *Mol Microbiol* 2006;**62**:480–90.

31. Morinaga N, Yahiro K, Matsuura G, Watanabe M, Nomura F, Moss J, et al. Two distinct cytotoxic activities of subtilase cytotoxin produced by shiga-toxigenic *Escherichia coli*. *Infect Immun* 2007;**75**:488–96.

32. Morinaga N, Yahiro K, Matsuura G, Moss J, Noda M. Subtilase cytotoxin, produced by Shiga-toxigenic *Escherichia coli*, transiently inhibits protein synthesis of Vero cells via degradation of BiP and induces cell cycle arrest at G1 by downregulation of cyclin D1. *Cell Microbiol* 2008;**10**:921–9.

33. Wolfson JJ, May KL, Thorpe CM, Jandhyala DM, Paton JC, Paton AW. Subtilase cytotoxin activates PERK, IRE1 and ATF6 endoplasmic reticulum stress-signalling pathways. *Cell Microbiol* 2008;**10**:1775–86.
34. Yamazaki H, Hiramatsu N, Hayakawa K, Tagawa Y, Okamura M, Ogata R, et al. Activation of the Akt-NF-kappaB pathway by subtilase cytotoxin through the ATF6 branch of the unfolded protein response. *J Immunol* 2009;**183**:1480–7.
35. Harama D, Koyama K, Mukai M, Shimokawa N, Miyata M, Nakamura Y, et al. A subcytotoxic dose of subtilase cytotoxin prevents lipopolysaccharide-induced inflammatory responses, depending on its capacity to induce the unfolded protein response. *J Immunol* 2009;**183**:1368–74.
36. Nakajima S, Hiramatsu N, Hayakawa K, Saito Y, Kato H, Huang T, et al. Selective abrogation of BiP/GRP78 blunts activation of NF-kappaB through the ATF6 branch of the UPR: involvement of C/EBPbeta and mTOR-dependent dephosphorylation of Akt. *Mol Cell Biol* 2011;**31**:1710–18.
37. Hayakawa K, Hiramatsu N, Okamura M, Yamazaki H, Nakajima S, Yao J, et al. Acquisition of anergy to proinflammatory cytokines in nonimmune cells through endoplasmic reticulum stress response: a mechanism for subsidence of inflammation. *J Immunol* 2009;**182**:1182–91.
38. Hayakawa K, Hiramatsu N, Okamura M, Yao J, Paton AW, Paton JC, et al. Blunted activation of NF-kappaB and NF-kappaB-dependent gene expression by geranylgeranylacetone: involvement of unfolded protein response. *Biochem Biophys Res Commun* 2008;**365**:47–53.
39. Okamura M, Takano Y, Hiramatsu N, Hayakawa K, Yao J, Paton AW, et al. Suppression of cytokine responses by indomethacin in podocytes: a mechanism through induction of unfolded protein response. *Am J Physiol Renal Physiol* 2008;**295**:F1495–1503.
40. Sato M, Yao VJ, Arap W, Pasqualini R. GRP78 signaling hub a receptor for targeted tumor therapy. *Adv Genet* 2010;**69**:97–114.
41. May KL, Paton JC, Paton AW. *Escherichia coli* subtilase cytotoxin induces apoptosis regulated by host Bcl-2 family proteins Bax/Bak. *Infect Immun* 2010;**78**:4691–6.
42. Yahiro K, Satoh M, Morinaga N, Tsutsuki H, Ogura K, Nagasawa S, et al. Identification of subtilase cytotoxin (SubAB) receptors whose signaling, in association with SubAB-induced BiP cleavage, is responsible for apoptosis in HeLa cells. *Infect Immun* 2011;**79**:617–27.
43. Yahiro K, Tsutsuki H, Ogura K, Nagasawa S, Moss J, Noda M. DAP1, a negative regulator of autophagy, controls SubAB-mediated apoptosis and autophagy. *Infect Immun* 2014;**82**:4899–908.
44. Xie WY, Zhou XD, Yang J, Chen LX, Ran DH. Inhibition of autophagy enhances heat-induced apoptosis in human non-small cell lung cancer cells through ER stress pathways. *Arch Biochem Biophys* 2016;**607**:55–66.
45. Misra UK, Pizzo SV. Receptor-recognized alpha(2)-macroglobulin binds to cell surface-associated GRP78 and activates mTORC1 and mTORC2 signaling in prostate cancer cells. *PLoS One* 2012;**7**:e51735.
46. Martin S, Hill DS, Paton JC, Paton AW, Birch-Machin MA, Lovat PE, et al. Targeting GRP78 to enhance melanoma cell death. *Pigment Cell Melanoma Res* 2010;**23**:675–82.
47. Backer JM, Krivoshein AV, Hamby CV, Pizzonia J, Gilbert KS, Ray YS, et al. Chaperone-targeting cytotoxin and endoplasmic reticulum stress-inducing drug synergize to kill cancer cells. *Neoplasia* 2009;**11**:1165–73.
48. Prabhu A, Sarcar B, Kahali S, Shan Y, Chinnaiyan P. Targeting the unfolded protein response in glioblastoma cells with the fusion protein EGF-SubA. *PLoS One* 2012;**7**:e52265.

49. Firczuk M, Gabrysiak M, Barankiewicz J, Domagala A, Nowis D, Kujawa M, et al. GRP78-targeting subtilase cytotoxin sensitizes cancer cells to photodynamic therapy. *Cell Death Dis* 2013;**4**:e741.

50. Gabrysiak M, Wachowska M, Barankiewicz J, Pilch Z, Ratajska A, Skrzypek E, et al. Low dose of GRP78-targeting subtilase cytotoxin improves the efficacy of photodynamic therapy in vivo. *Oncol Rep* 2016;**35**:3151–8.

51. Zhang L, Li Z, Shi T, La X, Li H, Li Z. Design, purification and assessment of GRP78 binding peptide-linked Subunit A of Subtilase cytotoxic for targeting cancer cells. *BMC Biotechnol* 2016;**16**:65.

52. Jindadamrongwech S, Thepparit C, Smith DR. Identification of GRP 78 (BiP) as a liver cell expressed receptor element for dengue virus serotype 2. *Arch Virol* 2004;**149**:915–27.

53. Duriez M, Rossignol JM, Sitterlin D. The hepatitis B virus precore protein is retrotransported from endoplasmic reticulum (ER) to cytosol through the ER-associated degradation pathway. *J Biol Chem* 2008;**283**:32352–60.

54. Wang L, Cheng W, Zhang Z. Respiratory syncytial virus infection accelerates lung fibrosis through the unfolded protein response in a bleomycin-induced pulmonary fibrosis animal model. *Mol Med Rep* 2017;**16**:310–16.

55. Buchkovich NJ, Maguire TG, Yu Y, Paton AW, Paton JC, Alwine JC. Human cytomegalovirus specifically controls the levels of the endoplasmic reticulum chaperone BiP/GRP78, which is required for virion assembly. *J Virol* 2008;**82**:31–9.

56. Wati S, Soo ML, Zilm P, Li P, Paton AW, Burrell CJ, et al. Dengue virus infection induces upregulation of GRP78, which acts to chaperone viral antigen production. *J Virol* 2009;**83**:12871–80.

57. Nain M, Mukherjee S, Karmakar SP, Paton AW, Paton JC, Abdin MZ, et al. GRP78 is an important host factor for Japanese Encephalitis virus entry and replication in mammalian cells. *J Virol* 2017;**91** e02274–16.

58. Zhang Y, Liu R, Ni M, Gill P, Lee AS. Cell surface relocalization of the endoplasmic reticulum chaperone and unfolded protein response regulator GRP78/BiP. *J Biol Chem* 2010;**285**:15065–75.

59. Misra UK, Pizzo SV. Ligation of cell surface GRP78 with antibody directed against the COOH-terminal domain of GRP78 suppresses Ras/MAPK and PI 3-kinase/AKT signaling while promoting caspase activation in human prostate cancer cells. *Cancer Biol Ther* 2010;**9**:142–52.

60. Gonzalez-Gronow M, Cuchacovich M, Llanos C, Urzua C, Gawdi G, Pizzo SV. Prostate cancer cell proliferation in vitro is modulated by antibodies against glucose-regulated protein 78 isolated from patient serum. *Cancer Res* 2006;**66**:11424–31.

61. Violin JD, DeWire SM, Yamashita D, Rominger DH, Nguyen L, Schiller K, et al. Selectively engaging beta-arrestins at the angiotensin II type 1 receptor reduces blood pressure and increases cardiac performance. *J Pharmacol Exp Ther* 2010;**335**:572–9.

62. Drake MT, Violin JD, Whalen EJ, Wisler JW, Shenoy SK, Lefkowitz RJ. beta-arrestin-biased agonism at the beta2-adrenergic receptor. *J Biol Chem* 2008;**283**:5669–76.

63. Violin JD, Lefkowitz RJ. Beta-arrestin-biased ligands at seven-transmembrane receptors. *Trends Pharmacol Sci* 2007;**28**:416–22.

64. Russo A, Soh UJ, Trejo J. Proteases display biased agonism at protease-activated receptors: location matters!. *Mol Interv* 2009;**9**:87–96.

INDEX

Note: Page numbers followed by "*f*" and "*t*" refer to figures and tables, respectively.

D

E

R

S

T

U

V

W

X

Z

Printed in the United States
By Bookmasters